MINDING YOUR BUSINESS WITH MAPINFO®

Angela Whitener and Jeff Davis

Minding Your Business with MapInfo

Angela Whitener and Jeff Davis

Published by:

OnWord Press
2530 Camino Entrada
Santa Fe, NM 87505-4835 USA

Carol Leyba, Publisher
David Talbott, Acquisitions Editor
Barbara Kohl, Associate Editor
Jean Cooksey, Development Editor
Cynthia Welch, Production Manager
Liz Bennie, Director of Marketing
Lauri Hogan, Marketing Services Manager
Laurel Avery, Production Editor
Lynne Egensteiner, Cover designer, Illustrator

All rights reserved. No part of this book may be reproduced or transmitted in any form or by any means, electronic or mechanical, including photocopying, recording, or by an information storage and retrieval system without written permission from the publisher, except for the inclusion of brief quotations in a review.

Copyright © Angela Whitener and Jeff Davis

First Edition, 1998

10 9 8 7 6 5 4 3 2 1

Printed in the United States of America

Library of Congress Cataloging-in-Publication Data

Whitener, Angela, 1961-

Minding your business with MapInfo / Angela Whitener and Jeff Davis.

       p.   cm.

Includes index.

1. Information storage and retrieval system--Business. 2. MapInfo Professional. 3. Geographic information systems. 4. Database management.
I. Davis, Jeff, 1957-   . II. Title.
HF5548.2.W4632   1998
650'.022'3--dc21   97-43201
                                                                           CIP

Trademarks

MapInfo, MapInfo Professional, MapBasic, MapMarker, and SpatialWare are registered trademarks of MapInfo Corporation. MapInfo Desktop, MapInfo SQL DataLink, MapDirect, and MapXtreme are trademarks of MapInfo Corporation. PRIZM and Claritas are registered trademarks of Claritas Inc. The cluster names, such as "Blue Blood Estates" and "Winner's Circle," are trademarks of Claritas Inc. Blue Marble Graphics and Geographic Tracker are registered trademarks. GeoTrack and GeoObjects are trademarks of Blue Marble. On Target Mapping, DRIVE, DRIVE PLUS, HazWasteInfo, ToxicReleaseInfo, and HurricaneInfo are trademarks of On Target Mapping. Charting Pro, ChartLink, MapChart, and Panel View are registered trademarks of MISA and MISC. Surround View is a trademark of MISA and MISC. Wal-Mart is a registered trademark of Wal-Mart Stores, Inc. ZIP Code and ZIP+4 are registered trademarks of the U.S. Postal Service. Microsoft and Windows are registered trademarks of Microsoft Corporation in the United States and other countries. All other terms mentioned in this book that are known to be trademarks or service marks have been appropriately capitalized. OnWord Press cannot attest to the accuracy of this information. Use of a term in this book should not be regarded as affecting the validity of any trademark or service mark.

Warning and Disclaimer

This book is designed to provide information on the use of MapInfo Professional software and other products and tools in the MapInfo product line. Every effort has been made to make the book as complete, accurate, and up to date as possible; however, no warranty or fitness is implied.

The information is provided on an *as is* basis. The authors and OnWord Press shall have neither liability nor responsibility to any person or entity with respect to any loss or damages in connection with or arising from the information contained in this book.

About the Authors

Angela Whitener is vice president of software development for IntelleVue, a MapInfo Strategic Partner located in Tulsa, Oklahoma. Formerly J. Davis & Associates, Inc., IntelleVue develops and markets customized MapInfo based applications to the retail petroleum, banking, political, and oil and gas exploration industries.

Prior to joining IntelleVue, Angela was a senior systems consultant for MPSI, Inc., in Tulsa and a software engineer for E-Systems in Dallas, Texas. A GIS software user and consultant since 1992, Angela co-authored *INSIDE MapInfo Professional* (OnWord Press, 1996) with Larry Daniel and Paula Loree; *Mapping with Microsoft Office* (OnWord Press, 1996) with Bill Creath; and *MapBasic Developer's Guide* (OnWord Press, 1997) with Breck Ryker. She holds a BS in mathematics and computer science from Oklahoma State University and an MS in human relations and business from Amber University.

Jeffrey Davis is founder and president of IntelleVue (formerly J. Davis & Associates, Inc.). Prior to launching JDA, Jeff spent 15 years with an industry leading software provider focused on site modeling and retail network planning for various industries. He holds a BS in business administration from the University of Tulsa. Before becoming IntelleVue, JDA received the 1997 MapInfo Authorized Partner of the Year Award.

Acknowledgments

I want to thank Jeff Davis, who tackled this writing project with me, as well as all of the MapInfo partners and clients who helped provide material without which this book would not have happened. Thanks to MapInfo Corporation, which is always there to support its partners, and which provided assistance with material for this book. And thanks to Jeff, Greg, Gary, Lee, and John of the IntelleVue team, who foster an atmosphere of teamwork and growth vital to growing our company and serving our clients.

A special thanks to the staff at High Mountain Press, who organized and published this project—especially to Jean Cooksey, who insisted we keep the project moving. Thanks to my family, who offered support and encouragement through each of my book writing projects. And thank you, Keith, my loving husband, who continues to encourage me to explore my potential.

—*Angela Whitener*

First, I would like to thank Angela Whitener for asking me to participate in the writing of this book. Without Angela's patience and guidance, my part would never have been finished. I would also like to thank Angela for her enthusiasm, professionalism, and commitment to our business, for without it we would have not achieved our successes.

Next, I want to thank all of the individuals who contributed case studies to the book. Many are clients, and I truly appreciate their participation, including Stan Nickel, Sandy Carter, Rick McDowell, and Kevin Sabin. I am also grateful to all of IntelleVue's clients for allowing us to provide mapping services and solutions that enabled the company to understand business uses of MapInfo.

A big thanks to the team at High Mountain Press, and particularly Jean Cooksey, who was fantastic to work with and who allowed me the freedom and time (especially around deadlines!) to complete my parts as I wished.

Most of all, I want to thank my family—my wife Karen, daughter Kelsey, and son Trent, as well as Larry, Cindy, Mom, and Dad—for all of the support and love, in this book writing venture and in everything I do.

—*Jeff Davis*

Contents

Introduction 1
    Mapping: A Persuasive Tool 1
    Visualize the Geographic Component in Business Data 9
    GIS and Business Mapping 12
    MapInfo: The Company 14
    MapInfo: The Product 16
    Book Audience 17
    Web Updates 18
    Typographical Conventions 18
        MapInfo Commands 18
    Summary 19

PART ONE: Business Applications of MapInfo 21

Chapter 1: Marketing, Advertising, and Sales 23
    Marketing and Advertising 24
        Lifestyle Segmentation Data 24
        Bank Application 28
        Billboard Advertising 31
        Spending Data 34
    Customer Service 36
        Thinking Ahead 40
    Site Selection and Analysis 41
        Demographics 41
        Competition 46
        Ring, Drive Distance, and Drive Time Trade Area Analysis .. 48
        Modeling 49
    Sales Territory Redistricting 52

Chapter 2: Decision Sciences 57

  Inventory Management 57
    Beverage Company Deliveries 58
    Supermarket Product Placement and Inventory 61
    Parking Lot Inventory 63
    Warehouse Inventory 63
  Transportation 66
    Railroads 66
    Roadways 73
    Mass Transit 75
  Global Positioning Systems 78
    Mobiletrack 82
    Circulation and Delivery 83
    Geographic Tracker 85
  Routing Logistics 88
    Tracking Utility Repairs 88
    Package Pickup and Delivery Solutions 89

Chapter 3: Telecommunications and Information Systems 95

  Telecommunications 95
    Marketing and Sales 97
    Customer Service 99
    Network Management and Planning 101
    Tracking Buried Cables 103
    Competitive Analysis 104
    Telecommunications Data 105
  Information Systems 105
    Intranet/Internet Applications 106
    Oil and Gas Well Mapping 108
    Accounting Management/Reporting 109
    Human Resources 110
    Politics 110
  Desktop Mapping in Brief 111

Chapter 4: Banking, Insurance, and Real Estate 113

  Banking 113

Contents

Mapping to the Rescue 114
Network Optimization/Site Selection 115
CRA Compliance 116
Marketing, Advertising, and Customer Analysis 118
Insurance 120
   Underwriting 120
   Risk Concentration Analysis 124
   Disaster Planning/Catastrophe Response 127
   Sales and Planning 128
Real Estate 131
   Site Selection 134
   Real Estate Management 135
   Improving the Home Search 137

PART TWO: Case Studies   139

Price Changes in Retail Petroleum Markets   143

Prototype Mission 144
   Tracking Customer Purchases 145
MapInfo Screens and Application Programming 146
   Site Map 148
   Pricing Analysis Menu 149
   Credit Card Customer Point Map 151
   Common Customers 153
   Credit Card Customer Point and Area Maps 154
   Parameter Choices 155
   Customer Migration 156
   Premium Payment Method Comparison 159
   Payment Method Comparison 160
Conclusion 161

Thrifty Rent-A-Car   163

Know Your Customers 164
Generating Maps 165
Emerging Trends 167
Conclusion 168

Rural Press and CDATA 171
    Using CDATA 173
    Successes with MapInfo 174
    Innovations in CDATA96 177

Canada Post 181
    Methodology 182
        Determine Customer Base 183
        Establish Store Attractiveness Factor 183
        Define Market Area 184
        Locate Existing Stores 184
        Determine Total Market Revenue Potential 184
        Create Grid of Store Attractiveness 185
        Establish Optimal Store Locations 189
        Perform Gap Analysis with Field Personnel 189
        Generate Patronage Probability Grid 189
        Determine Relative Store Strength 191
        Determine Proposed Store Revenue 192
    Conclusion 192

Emergency Medical Service Institute 193
    Response Time 195
    Licensing 198
    Conclusion 198

PageNet 201
    Defining the Problem 202
    Solving the Problem 202
        Launching PCS SiteManager 203
        Marketing/Executive Level Overview Module 205
        Site Reporter Module 206
        Real Time Site Analysis Probe 207
    Future Enhancements 210
        Obstacle Module 210
        Site Comparator Module 210

Louis Dreyfus Natural Gas   213

Problem Definition 215
9-section Plat Maps 216
    Data Files 218
Creating the Map 221
    Oil and Gas Well Data 222
    Identifying a Target Section 224
Other Applications 227

Visimark   229

Versatility 229
Searching and Database Capabilities 231
Saving Time, Money, and Drive Space 233
GIS Interface 233
    Geocoding 235
    Searching by Map Location 236
    Mapping a Selected Set of Properties 236

Appendix A: Glossary   239

Appendix B: Reference Material and Data Sources   247

OnWord Press MapInfo Books 247
E-mail List 248
Satellite and Aerial Data Sources 248
    Remote Sensing Resources on the Web 249
Telecom Data 249
Bureau of Transportation Statistics Data 252
Internet Sites 253

Appendix C: MapInfo Product Line Overview   271

Product Information 273

Appendix D: Contact Information   281

Index   287

INTRODUCTION

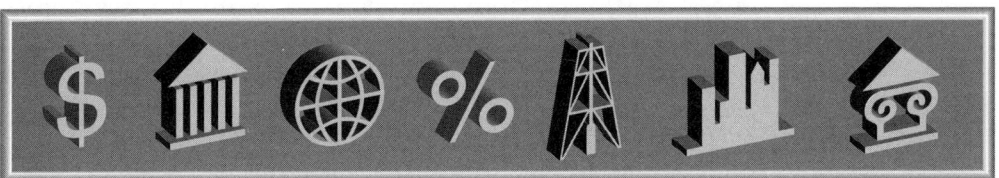

Mapping: A Persuasive Tool

The adage "a picture is worth a thousand words" certainly holds true in the context of mapping business data. Combining tabular data with charts or maps makes trends easier to detect and analyze, and maps and charts are powerful tools to communicate results to others.

Over time, electronic storage of business data has increased exponentially, and today business data are typically stored in massive corporate databases. As data have grown, so has the need to analyze them with greater speed. Spreadsheets and graphs have enhanced the ability to analyze data in recent years, quickly becoming part of every business PC's desktop. Using mapping technology in conjunction with such tools can revolutionize data analysis and presentation.

A simple example demonstrates the effectiveness of mapping to visualize business data. Every 10 years the U.S. government conducts a census of the country's inhabitants, and the statistical data can be summarized by state. The following discussion is focused on addresses state name, population, number of households, and income data. Summary census information viewed on a spreadsheet is depicted as rows and columns of data.

	A	B	D	F	G	B
1	State_Name	State	Pop_1990	Num_Hh_90	Med_Inc_80	
2	Alabama	AL	4040587	1506790	21714	
3	Alaska	AK	550043	188915	34130	
4	Arizona	AZ	3665228	1368843	25218	
5	Arkansas	AR	2350725	891179	20361	
6	California	CA	29760021	10381206	33342	
7	Colorado	CO	3294394	1282489	28558	
8	Connecticut	CT	3287116	1230479	36961	
9	Delaware	DE	666168	247497	28887	
10	District Of Co	DC	606900	249634	26962	
11	Florida	FL	12937926	5134869	25914	
12	Georgia	GA	6478216	2366615	26342	
13	Hawaii	HI	1108229	356267	34997	
14	Idaho	ID	1006749	360723	24475	
15	Illinois	IL	11430602	4202240	31119	

Spreadsheet of census summary data.

Spreadsheet software can create charts and graphs to show relationships among data. For example, the following graph shows median income for 13 states. The graph indicates that New Jersey has a high median income compared with states such as Oklahoma and Mississippi.

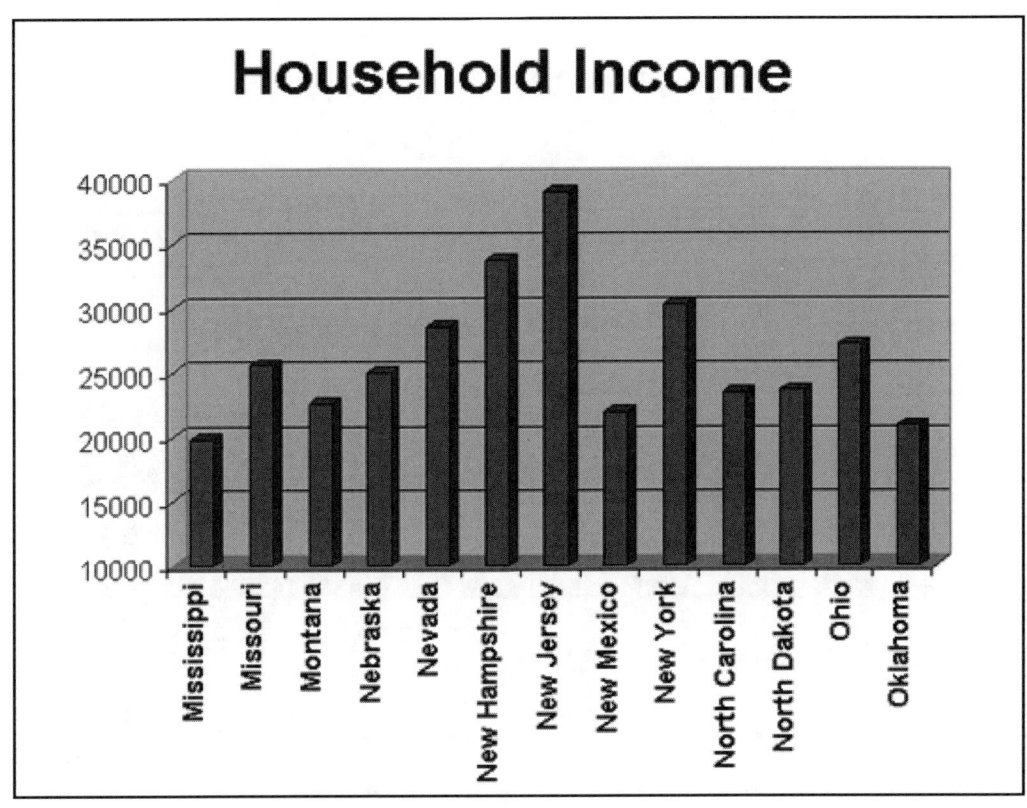

Graph of selected states by median income.

However, when the graph is expanded to include all states it becomes far more difficult to interpret.

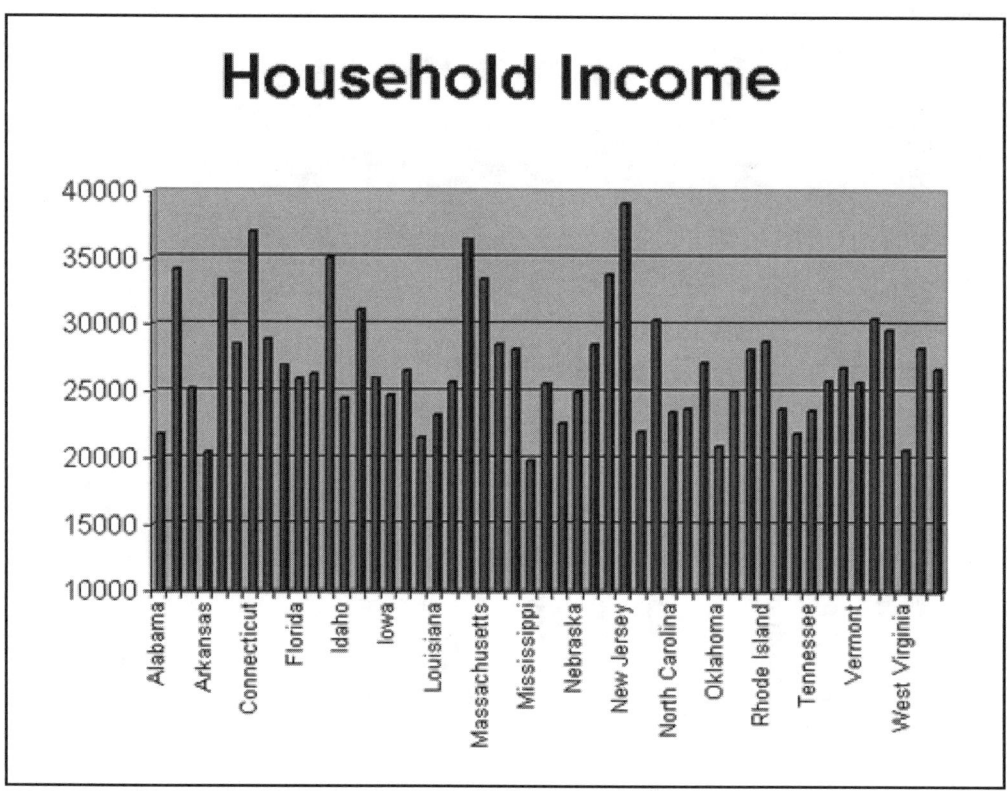

Graph of all states by median income.

The above example illustrates the limits of visualizing data on a graph—depicting 50 variables in this manner is inappropriate for analysis. The graph should be wider to accommodate state names along the X axis. But a wider graph also necessitates paging back and forth to view all the data. However, it is possible to effectively display median income data on a map. The following map color codes each state according to median income range for that state. As a result, the map clearly shows high and low income areas relative to one another (in grayscale, darker tones indicate higher median income values).

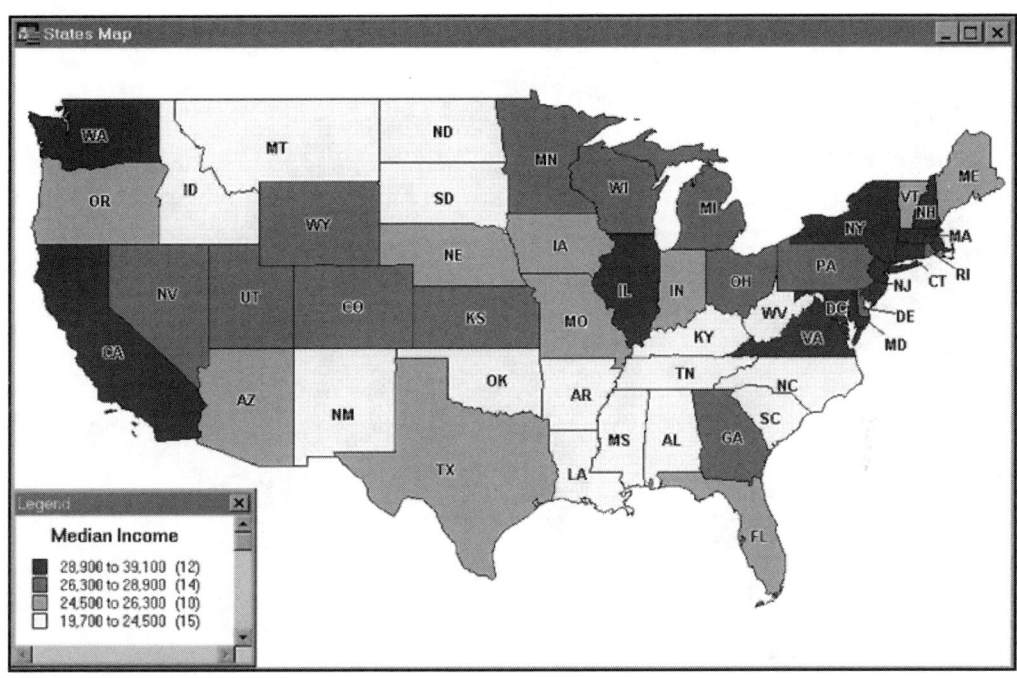

U.S. map showing median income.

MapInfo enables users to create a virtually limitless range of maps, many of which are explored in this book. The next example shows a U.S. map organized according to number of households per state. In addition, a thematic layer of pie charts shows the relationship between urban and rural populations in each state.

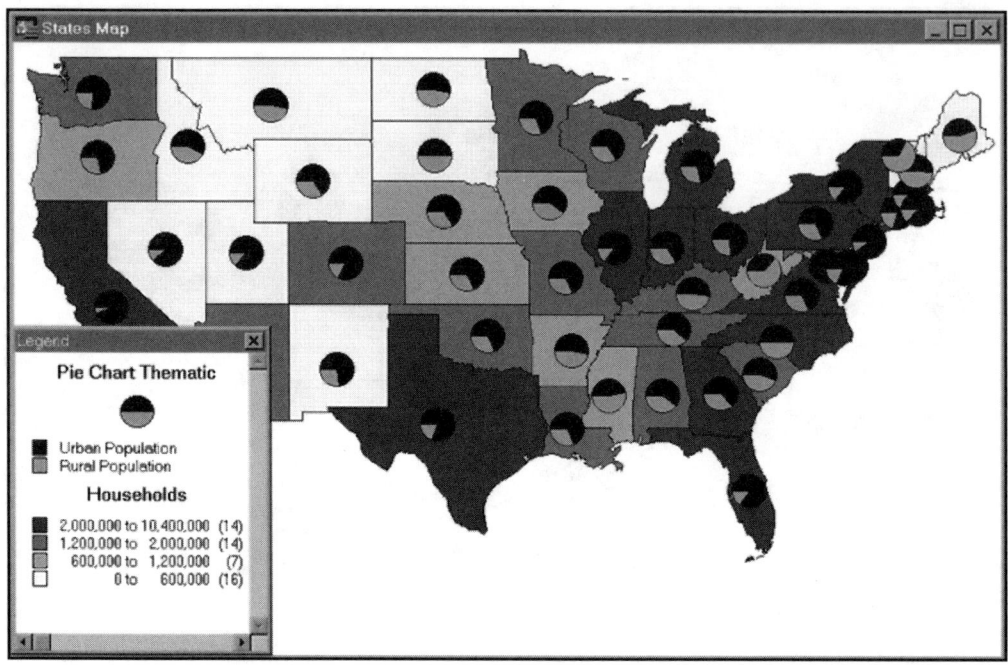

U.S. map showing number of households, and proportions of urban and rural population.

While some would say this is an impressive amount of information to display on a single map, the functionality of business mapping is actually far richer. If you find a map becoming congested—or the scope of your research narrows—you can zoom in for a more detailed analysis. The preceding example depicts the entire United States. For more detail you can zoom in on the northeastern United States, as shown in the next illustration.

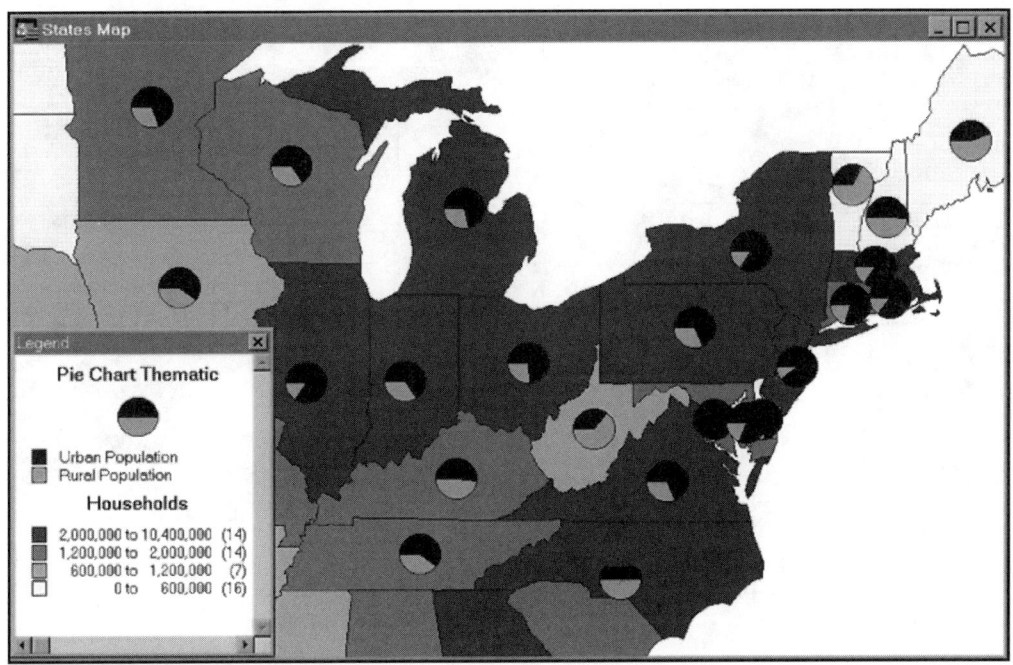

Zoom in to see greater detail.

With MapInfo, you can also adjust how you view a map and add information. The next figure shows the same information as the preceding map but depicts the south central United States. In addition, a layer of information containing labels for major cities and interstate highways has been added.

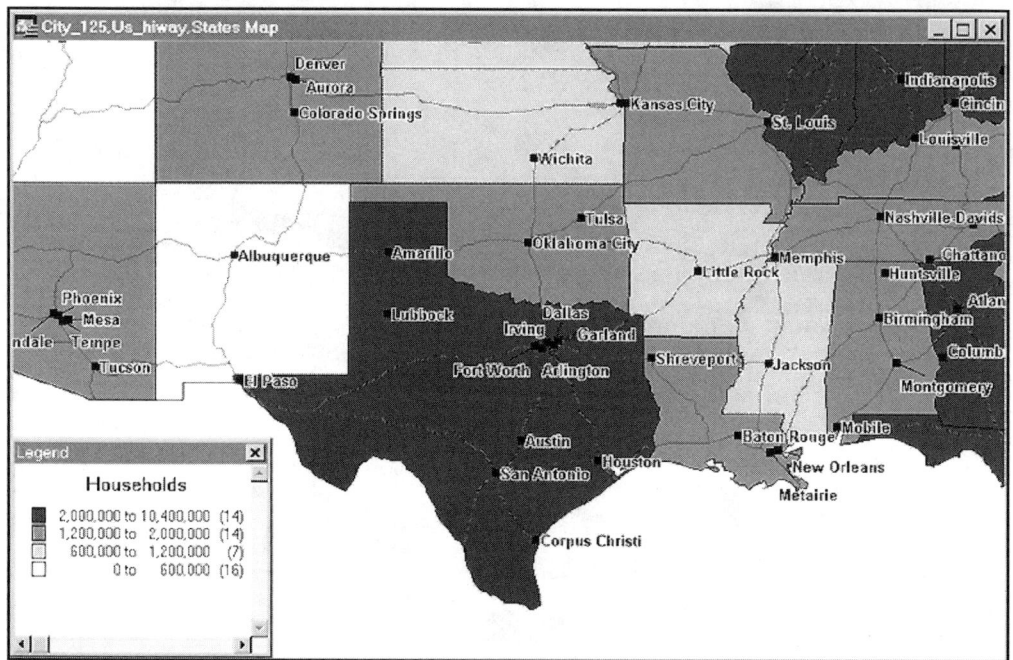

Map of the south central United States showing number of households, major cities, and interstate highways.

As you become familiar with a data set, you can make inferences and conclusions regarding its content. However, it can be difficult to communicate such information to others. Yet this ability is vital in business, particularly if your goal is to quickly and persuasively convey your conclusions to company executives. Maps can often help you accomplish this objective.

Visualize the Geographic Component in Business Data

Large volumes of business information are available today, and the amount grows daily. (For a sample of available data, refer to the extensive tables in Appendix B, "Reference Material and Data Sources.") Current and historical information exists in spreadsheets, marketing records, sales records, and other forms. Many professionals use such information to analyze trends and make business decisions. Data contain a wealth of information about customers, stores, personnel, equipment, and resources.

Nearly all current business data have a geographic component. It has been estimated that 85 percent of all databases contain some form of geographic information—street addresses, city or state names, ZIP Codes, or telephone numbers with area codes—almost all of which can be summarized and visualized on a map.

What does the term *geographic reference* mean? Whether you deal with employees, suppliers, or customers, you will likely have at least one database containing information similar to the data depicted in the next image.

First_Name	Address	City	State	Zip
MIKE/JOY	5021 SW HOLIDAY	CLAREMORE	OK	74017
PATRICIA	612 S MAGNOLIA AV	BROKEN ARROW	OK	74012
AMY/DAVID	5678 S UTICA	TULSA	OK	74105
DEBRA	2201 W LOUISVILLE	BROKEN ARROW	OK	74012
LAURA ANN	7320 E MARSHALL PL	TULSA	OK	74115
RANDY	8009 S 86 E AV	TULSA	OK	74135
JOHN	1308 S LOUISVILLE	TULSA	OK	74112
SAMUEL	3737 E VIRGIN PL	TULSA	OK	74115
JEANNE	RT 1 BOX 306	INOLA	OK	74036

Customer database records.

In the above database, each record has an address field that can be used to determine the location (or point) of a customer. You can use point information to analyze where a store is located and where customers live. The database also contains information that connects customers to geographic regions (e.g., city, county, and state) that are easy to understand and convey. Regional data can often supplement customer data, allowing you to summarize information for broader analysis, such as breaking down customer shipments by state. Some databases contain city, state, or ZIP Code information if not actual customer addresses. In the next example, customer records can be located as city points, the center of ZIP Code regions, or center of a state. Telephone area codes are yet another example of frequently overlooked geographically referenced data.

Lname	Fname	Company	City	State	Zip	Order_amt
Urdahl	Mike	GOLDMAN SACHS & COMPANY	Columbia	MD	21046	13,880
Olsen	Kevin	CHERMAN COMPANY	Dallas	TX	75225	13,780
Berger	S D	MERCEDES & CO.	Hermosa Beach	CA	90254	16,980
Parker	Al William	STATE PERSONNEL DEPARTMEN	Freeport	LA	71105	7,600
Taylor	Brian	COLORADO DEPT. OF HIGHWAYS	Columbus	OH	43210	6,240
Wang	Darrel J	DATA TERMINAL SERVICES, INC.	Lawndale	CA	90260	16,880
Berry	Ed	DARLING SCHOOL DISTRICT	Monroe	LA	71203	4,280
Hunsaker	R J	BEATRICE	Collingswood	NJ	08108	10,320
Swanson	Mike	CITY OF AKRON	Brick	NJ	08724	8,420
Patton	Hugo	AMARILLO COLLEGE	New York	NY	10006	8,220
Tracy	Rob	GOLDMAN SACHS & COMPANY	Ithaca	NY	14850	13,260
Boileau Jr	Marcia	WILSON SNYDER PUMPS	Gaithersburg	MD	20879	14,500
Caton	Joseph	US NAVY	Baltimore	MD	21227	13,640
Williams	Jason	VI SOFTWARE DEVELOPMENT	Minster	OH	45865	9,000
Genekens	Robert	ALL WEST COMPUTER	Baton Rouge	LA	70816	2,240
Drazkowski	James	D & M AIR COURIER	Campbell	CA	95008	4,100

Customer database records suitable for analysis by state or ZIP Code.

The level of geographic detail you use for analysis depends on the type of analysis you perform and the results you seek. While owners of a fast food restaurant want to analyze local address records to see where customers originate, national election analysts are likely to focus on voter precincts summarized regionally or by state. Other geographic references include counties, census block groups, school districts, and sales territories.

MapInfo can help you sort such information and display it geographically. Mapping enables you to quickly perceive patterns and relationships in large amounts of data. MapInfo displays data as points or thematic shaded maps, pie charts, and bar graphs. In addition, you can add labels, text, and formatted data to a map to enhance presentation.

GIS and Business Mapping

A geographic information system (GIS) uses automated mapping and database technology to depict data on digital maps. GIS technology allows you to create, store, maintain, retrieve, analyze, and display a variety of geographic and tabular information. The next illustration shows how a GIS might depict the state of Colorado and associated data.

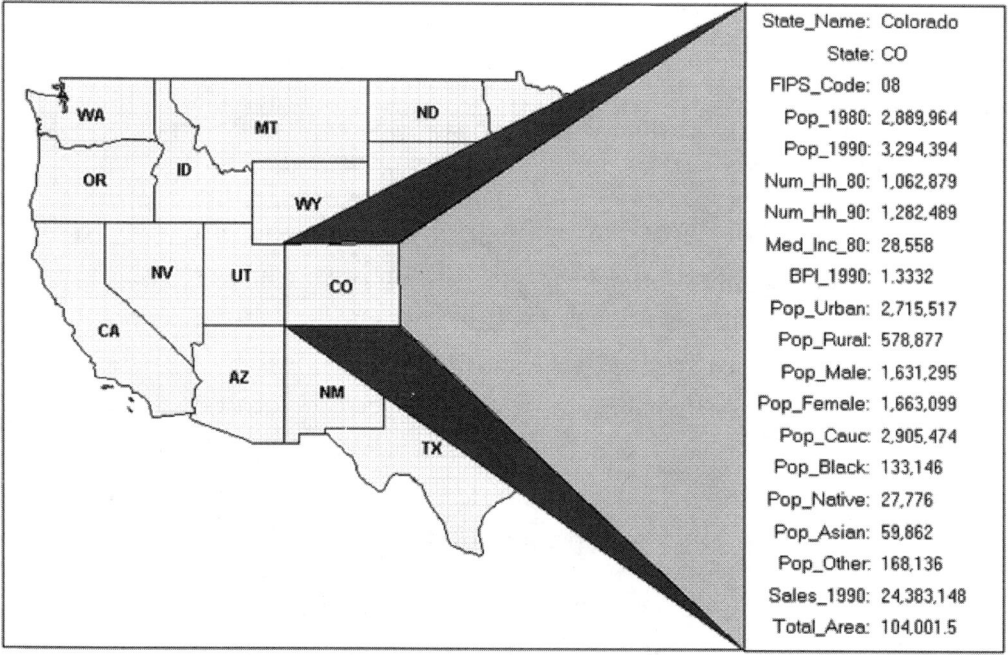

Feature data (state map at left) and attribute data (table at right) working together.

A GIS record file describing the state of Colorado includes fields of text and numeric information, as well as fields of spatial data, enabling the computer to draw the state as a region of specific size and shape. Think of a GIS as a database that stores location and shape information (*feature data*) as well as text and numeric information (*attribute data*).

GIS has the capability to store and access logically connected layers of data in any order or combination. In a GIS environment, the three layers below are positioned together and viewed as they geographically relate to one another.

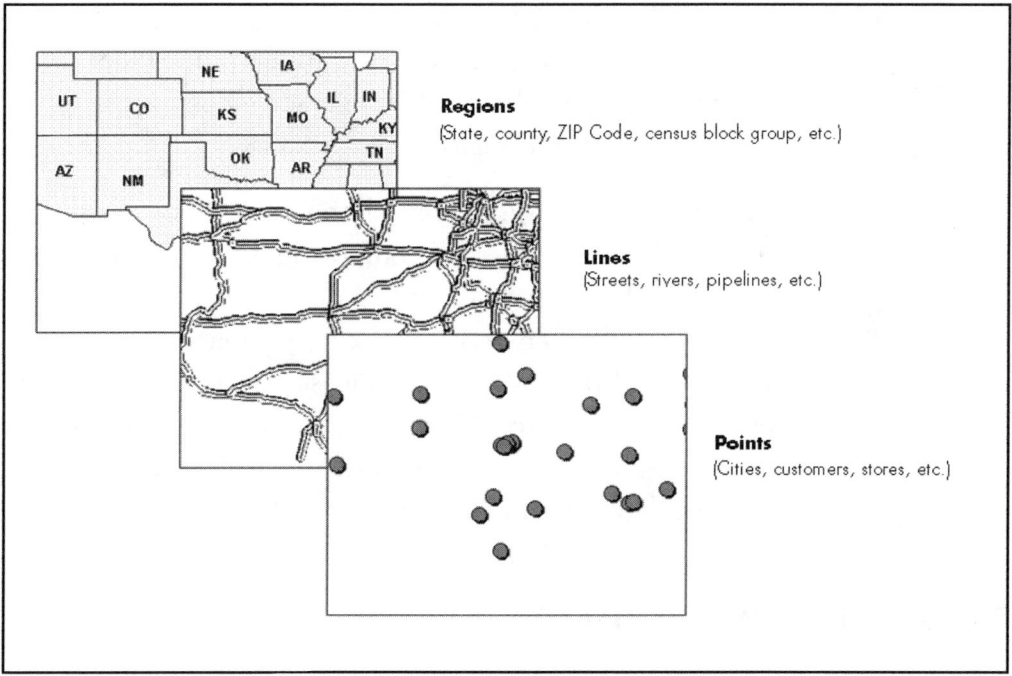

Three potential GIS layers.

While GIS is sometimes viewed as a computer system for drawing maps, in actuality the reverse is true: *GIS is* intended to map and analyze how data are distributed geographically. With a GIS, you view data in the form of a map (not a spreadsheet), which has a substantial impact on the conclusions you can draw from data.

In the early days of computerized mapping and GIS, expensive computer hardware and processing power were the norm. However, in the late 1980s PC hardware processing power had increased and cost had declined enough that the average business computer user could access systems sufficient for working with maps. Next, the release of the 1990 census data and TIGER (Topologically Integrated Geographic Encoding and Referencing) files developed by the U.S. Census Bureau made electronic mapping data affordable and relatively accessible.

MapInfo Corporation coined the term *desktop mapping* within the evolving PC computing environment. Now known as business mapping, the market niche is a component of the larger branch of information processing known as GIS. With the release of its product, now called MapInfo Professional, the company deployed GIS technology that enables users to visualize, analyze, and present data on a PC platform.

MapInfo: The Company

MapInfo Corporation, the worldwide leader in business mapping solutions, offers a family of software and data products to help businesses interpret data, conduct spatial analyses, and improve decision making through the visual display of data.

Founded in 1986 by four students at Rensselaer Polytechnic Institute, MapInfo Corporation is headquartered in Troy, New York. At present, MapInfo offers GIS mapping solutions in at least 20 languages and 58 countries. The number of users exceeds 200,000, and platforms include standalone desktops, client/server networks, data warehouses, and the Internet. In February 1994, MapInfo went public and can now be found on NASDAQ under the symbol MAPS.

In recent years MapInfo has formed alliances with Informix Software, Inc., Oracle Corporation, and Microsoft Corporation, among other companies, to integrate spatial and mapping technology into their products. Decision support companies such as Pilot Software, Inc., Andyne Computing Limited, and Information Advantage, Inc., are also integrating MapInfo technology into their solutions.

The partnership with Microsoft is particularly noteworthy. Beginning with Microsoft Office 95, Microsoft added a new mapping feature that can be used to create a map based on Excel spreadsheet data. The feature makes the spatial dimension of Microsoft Office 95 data accessible and brings the relationships between the spatial dimension and other variables into focus.

MapInfo has also led efforts to reduce the time and expertise required to generate maps, taking the technology from the domain of the specialist to less technically proficient users, which will make spatial analysis more accessible than ever before.

MapInfo: The Product

MapInfo Professional users can be found in business settings where problems such as site selection, distribution and routing, asset and network management, marketing, and customer service are key functions. The following list describes typical users.

- Market researchers
- Sales managers
- Planning officers
- Customer service representatives
- Risk analysts
- Network engineers
- Dispatchers
- Business executives
- Law enforcement investigators

Because MapInfo technology is very flexible or horizontal in nature (i.e., it can be applied in virtually any industry), it has been effective across a wide variety of markets. Typical markets for MapInfo customers follow.

- Telecommunications
- Retail
- Banking
- Insurance
- Health care
- Real estate

- Utilities
- Government
- Transportation and delivery

Early on, because the company recognized that it could not adequately serve the fast growing mapping market, it created a team of integrators and value added resellers (such as IntelleVue). Teaming with individuals and companies who understand current and future business challenges is a key element of the software's success. The company's network of partners works with the MapInfo product line to supplement functionality and customize it for a variety of business uses.

Book Audience

This book is aimed at professionals with a basic knowledge of mapping software and individuals responsible for business analysis. All such individuals can benefit from reading the text. Proficiency in MapInfo is not a necessity but it can decrease the learning curve when applying the techniques discussed. Part One offers an overview of various business applications of MapInfo Professional and third party products. Part Two presents several case studies describing real world MapInfo applications.

Web Updates

Readers curious about updates or corrections to this book may access the Web site *www.onwordpress.com/updates.cfm*.

Typographical Conventions

The following shorthand conventions are used in this book to identify distance measurements: 30" (30 inches); 25' (25 feet); and 6.2 mi (6.2 miles). Metric distances are abbreviated as follows: 2.5m, 130km, 0.3cm. All measurements are approximate unless otherwise noted.

Web addresses and other information that a MapInfo user must type (see the next section) are *italicized*.

MapInfo Commands

The names of MapInfo user interface items such as menus, windows, menu items, tools, toolbars, icons, and dialog box options are capitalized. Next, command sequences (clicking on one or more functions in succession) are separated by a pipe (|). Examples follow.

- Using the Frame tool, draw three frames for a map, browser, and graph. Assign None as the window linked to each frame.
- Choose Query | SQL Select on the main menu bar.

User input and names for files, directories, tables, fields, column names, and so forth are *italicized*. Examples follow.

- Set the Graph dialog box to graph the *Pop_1980* and *Pop_1990* columns of the table.
- Open a Browser window of the *STATES* table.

Summary

This chapter introduced the concept of business mapping and illustrated how maps can display geographically referenced data that many businesses encounter on a daily basis. Displaying data on a map facilitates analysis, and combining maps with other business data significantly increases analytical capabilities.

Part One of the book provides more information regarding how mapping and GIS techniques have been used to solve various business problems. Part Two presents case studies that illustrate real world MapInfo applications.

PART ONE

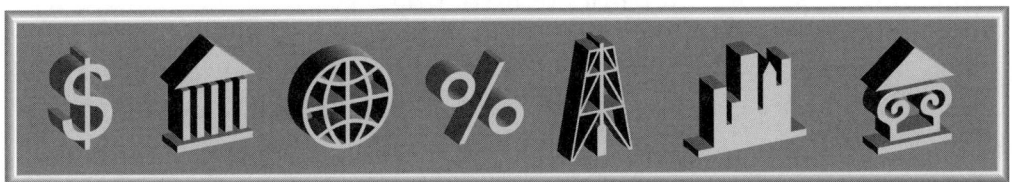

BUSINESS APPLICATIONS OF MAPINFO

Applications of MapInfo Professional are rapidly increasing as more industries learn to incorporate mapping techniques into decision making and analysis. This section outlines industries and business activities that rely on desktop mapping, and particularly MapInfo, to gain competitive advantage, understand and serve customers more efficiently, and comprehend the geographic implications of a customer base.

- **Chapter 1** focuses on the versatility of desktop mapping software with regard to advertising, marketing, customer service, site selection, and sales territory redistricting.

- **Chapter 2** explains how desktop mapping is being used in the field of decision sciences to support routing and warehousing. A discussion of global positioning systems is included.

- **Chapter 3** describes the uses of desktop mapping in telecommunications and information systems.

- **Chapter 4** explains how the banking, insurance, and real estate industries benefit from mapping techniques.

1

Marketing, Advertising, and Sales

The broad category of marketing likely represents the widest application of MapInfo products because it involves sorting volumes of geographically referenced data. In every industry, businesses that incorporate desktop mapping solutions into marketing, advertising, and sales efforts can benefit from easier data access and sharing, more informed and timely decision making, increased productivity, and better use of resources. This chapter describes how MapInfo can be applied in the following activities.

- Marketing and advertising
- Customer service
- Site selection and analysis
- Sales territory redistricting

However, because the marketing applications of MapInfo are so widespread, additional applications can be found elsewhere in Part One.

Marketing and Advertising

A substantial component of most business endeavors, marketing can at times account for 50 cents or more of every consumer dollar. Today, mapping technology is primarily used to assist in market segmentation and advertising. Also known as target marketing, segmentation concentrates efforts on customers within a market most likely to be interested in a given product or service, rather than the market as a whole. Markets can be segmented according to geographic, demographic, and other variables.

Lifestyle Segmentation Data

Several business data providers classify U.S. census (or demographic) data according to coherent lifestyle patterns. One such system, PRIZM, is offered by Claritas Inc. PRIZM segments each U.S. neighborhood into clusters based on similar demographic characteristics. The approach reflects the adage that "birds of a feather flock together," or that people gravitate toward neighborhoods and neighbors compatible with their lifestyles. Consequently, grouping marketing efforts according to neighborhoods is efficient because residents of the same neighborhood are more likely to respond to similar products. The table below lists the lifestyle segmentation clusters that constitute the PRIZM system.

Marketing and Advertising

Social group	Cluster number	Nickname	Demographic caption
Elite suburbs	1	Blue Blood Estates	Elite super-rich families
	2	Winner's Circle	Executive suburban families
	3	Executive Suites	Upscale white collar couples
	4	Pools & Patios	Established empty nesters
	5	Kids & Cul-de-Sacs	Upscale suburban families
Urban uptown	6	Urban Gold Coast	Elite urban singles and couples
	7	Money & Brains	Sophisticated townhouse couples
	8	Young Literati	Upscale urban singles and couples
	9	American Dreams	Established urban immigrant families
	10	Bohemian Mix	Bohemian singles and couples
Second city society	11	Second City Elite	Upscale executive families
	12	Upward Bound	Young upscale white collar families
	13	Gray Power	Affluent retirees in Sunbelt cities
Landed gentry	14	Country Squires	Elite exurban families
	15	God's Country	Executive exurban families
	16	Big Fish, Small Pond	Small town executive families
	17	Greenbelt Families	Young, middle class town families
The affluentials	18	Young Influentials	Upwardly mobile singles and couples
	19	New Empty Nests	Upscale suburban fringe couples
	20	Boomers & Babies	Young white collar suburban families
	21	Suburban Sprawl	Young suburban townhouse couples
	22	Blue-Chip Blues	Upscale blue collar families

Social group	Cluster number	Nickname	Demographic caption
Inner suburbs	23	Upstarts & Seniors	Middle income empty nesters
	24	New Beginnings	Young mobile city singles
	25	Mobility Blues	Young blue collar/service families
	26	Gray Collars	Aging couples in inner suburbs
Urban mid-scale	27	Urban Achievers	Mid-level, white collar, urban couples
	28	Big City Blend	Middle income immigrant families
	29	Old Yankee Rows	Empty nest, middle class families
	30	Mid-City Mix	African American singles and families
	31	Latino America	Hispanic middle class families
Second city centers	32	Middleburg Managers	Mid-level white collar couples
	33	Boomtown Singles	Middle income young singles
	34	Starter Families	Young middle class families
	35	Sunset City Blues	Empty nests in aging industrial cities
	36	Town & Gowns	College town singles
Exurban blues	37	New Homesteaders	Young middle class families
	38	Middle America	Mid-scale families in mid-sized towns
	39	Red, White, and Blue	Small town, blue collar families
	40	Military Quarters	GIs and surrounding off-base families
Country families	41	Big Sky Families	Mid-scale couples, kids and farmland
	42	New Eco-topia	Rural white/blue collar/farm families
	43	River City, USA	Middle class rural families
	44	Shotguns & Pickups	Rural blue collar workers and families

Marketing and Advertising

Social group	Cluster number	Nickname	Demographic caption
Urban cores	45	Single City Blues	Ethnically mixed urban singles
	46	Hispanic Mix	Urban Hispanic singles and families
	47	Inner Cities	Inner city, solo parent families
Second city blues	48	Smalltown Downtown	Older renters and young families
	49	Hometown Retired	Low income, older singles and couples
	50	Family Scramble	Low income Hispanic families
	51	Southside City	African American service workers
Working towns	52	Golden Ponds	Retirement town seniors
	53	Rural Industria	Low income, blue collar families
	54	Norma Rae-ville	Young families, biracial mill towns
	55	Mines & Mills	Older families, mine and mill towns
Heart-landers	56	Agri-business	Rural farm town and ranch families
	57	Grain Belt	Farm owners and tenants
Rustic living	58	Blue Highways	Middle income blue collar/farm families
	59	Rustic Elders	Low income, older, rural couples
	60	Back Country Folks	Remote rural/town families
	61	Scrub Pine Flats	Older African American farm families
	62	Hard Scrabble	Older families in poor isolated areas

The following illustration shows a census tract polygon and the PRIZM lifestyle clusters associated with it. The census tract shows a total population of 3,256 people distributed across three lifestyle clusters—1,236 individuals fall in the Urban Gold Coast cluster, 1,579 in Money & Brains, and 441 in Urban Achievers.

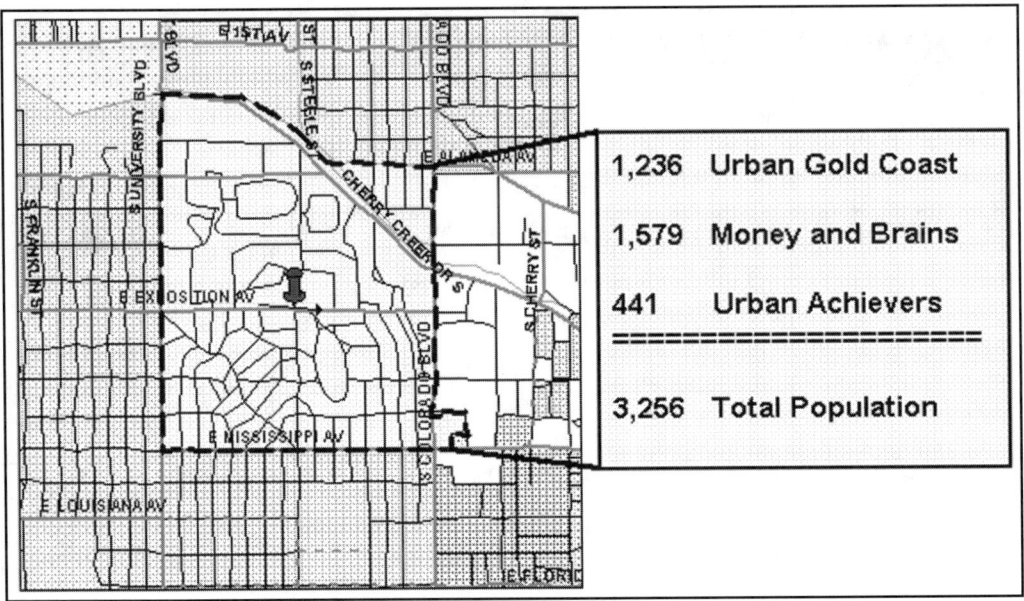

Census tract with associated lifestyle clusters.

If a business opens at the center of the above map (the pushpin), lifestyle knowledge would help marketers determine the customer base—clusters that presumably identify intelligent, wealthy people with style. Consequently, the products, decor, and marketing for a store in this census tract would differ greatly from one dominated by Shotguns & Pickups households. Adding lifestyle profiles to traditional demographic analysis offers more detailed market information and enables professionals to tailor advertising to specific audiences.

Bank Application

Lifestyle segmentation data can provide substantial insight into the customer's world. For example, bank X has designed a comprehensive business geographics application to profile customers using lifestyle clusters and other demographic databases, in order to assist in site selection, facility design, and product target marketing. Quite simply, identifying the "lifestyle" of a customer base allows marketers to make better decisions regarding how to best meet customer needs and maintain the relationship.

Site Selection and Facility Mix

When the bank contemplates opening a new branch, lifestyle segmentation data will help it define and serve its existing customer base. That is, building a full service branch with full staff and drive-through service would be appropriate if the lifestyle segmentation data indicate that the base is largely composed of older, affluent customers with multiple accounts and investments, and who favor traditional banking practices. However, a limited branch with drive-through service makes more sense if the data indicate a customer base composed of lower income, blue collar families with children and no more than two accounts. Such customers are likely to visit the branch less often and may appreciate the convenience of a drive-through because time is at a premium.

In conjunction with other demographic data, lifestyle segmentation may also help identify habits that dramatically alter how bankers serve the customer base. With the advent of electronic banking and the proliferation of Internet usage, certain customers may not require a local banking facility at all. These "electronic" customers can have all banking needs satisfied using the Internet and certain well-placed automatic teller machines (ATMs). The ability to minimize investments in facilities and personnel while serving the customer base directly impacts the bank's bottom line and profitability. Being able to identify and understand customer lifestyle trends—and therefore their banking needs—can only help bank managers make decisions.

Product Target Marketing

Lifestyle segmentation information goes hand in hand with product target marketing. However, while necessary, advertising is also expensive. Most advertising campaigns are broad in approach: they broadcast a message to many consumers and hope it reaches people who will respond. Today more and more firms are emphasizing target marketing to lower costs and improve sales success rates. If the bank plans to initiate a mutual fund campaign, it makes little sense to advertise to customers whose financial position limits their investment options. A bank can develop profiles of typical mutual fund customers by analyzing its existing customer base.

Once optimal potential customers are identified and collected in a database, MapInfo can geocode the records and map them for analysis. Trends or clusters will appear geographically and provide target areas for advertising and mailings. Recall that lifestyle segmentation is based on the assumption that people who live near one another share certain economic and social characteristics. Consequently, when existing mutual fund customers are located, the bank can obtain mailing lists for new customers in the same area.

Billboard Advertising

In another example of marketing, the company MISA combines GIS and MapInfo to display and maintain billboard inventory. Three levels of mapping applications offer various levels of sophistication: using MapInfo as a base, ChartLink offers simple display maps; MapChart offers a two-way mapping interface with the database program; and Outdoor DemoTrack combines census demographics and proprietary traffic flow studies to thematically map areas. In conjunction with Charting Pro software, these mapping applications allow users to produce maps showing billboard locations for proposed and/or contracted showings. Customized commands allow users to display contract, advertiser, and/or location data from the Charting Pro database. Potential posting line of sight conflicts and site restrictions are automatically checked before a board can be changed, and the changes are displayed in MapInfo. Add-on software such as Panel View and Surround View multimedia programs offer virtual "like" views of selected billboard sites using photographs, and other add-on software performs proximity buffering around a location using yellow page listings on compact disk.

Geocoding in MapInfo

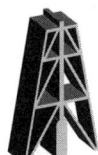

Street addresses are the most common form of geographic data. Address geocoding in MapInfo allows you to create unique map points that correspond with database addresses and carry accurate geographic positions. Each point connects a geocoded street file (provided with the software) with an address in a data file. A 100 percent match of street addresses and geocoded street file data—which produces a geocoded "layer" in MapInfo terminology—occurs only when the data and street files precisely match.

Because of the variety of addressing conventions and the potential for error, a 100 percent match is rare. However, there are ways to improve your chances. First, ensure that address data are accurately coded and that address prefixes and suffixes are coded to match the format used in the street network. For example, "1400 N. 16th Ave" may provide a match while "1400 N. 16th Av" may not.

Third party software packages can enhance the success rate as well. Such software "scrubs" data file addresses, cleaning street names for consistency with standards established by the U.S. Postal Service. Even clean keyboard entry of addresses will not provide a match if a street is incorrectly coded at the outset. To ensure a high percentage of matches, verify that the geocoded street network is accurate and current.

Raw TIGER/Line street files from the U.S. Census Bureau are likely to contain errors and omissions. While errors do not render TIGER files unusable, they can certainly reduce the "hit" rate when geocoding. You can edit TIGER files to improve accuracy but purchasing a revised street network from a commercial data provider may save you time and resources.

Some companies also provide address matching services and software for purchase. For example, the MapInfo product MapMarker produces acceptable geocoding hit rates. MapMarker assigns latitude and longitude at the street, ZIP Code, ZIP+2, and ZIP+4 levels in a single file pass. It can locate virtually every address and place name in the United States. Updated bi-monthly, MapMarker geocodes large files quickly and accurately and features a user friendly interface. Imbedded in MapMarker are sophisticated matching rules, derived from the most recent U.S. Census and Post Office data, to find and match addresses containing misspellings, omissions, and incorrect data.

International data files introduce different addressing issues. MapInfo abbreviation files must take into account different language strings and abbreviations (as in some European countries), or separate geocoding modules must be created altogether (to handle the Japanese addressing system, for example). Beginning with the release of version 3.0, MapInfo can accommodate "reversed" address orders (e.g., European and other addressing systems). The version of MapInfo released in Japan contains a geocoding module specifically for the Japanese addressing system.

Spending Data

In addition to lifestyle segmentation, which relies on census data, marketers also target efforts according to purchasing data gathered through credit card receipts, business accounts, and other means. These data can be geocoded and mapped, allowing a marketer to view customer distribution in relation to the product or service offered.

For example, one major oil company tracked credit card customers by their purchases to identify several issues, including trade area size and buying trends (e.g., average transaction data and crossover incidents). For more information, see the "Price Changes in Retail Petroleum Markets" case study in Part Two.

While enlightening, tracking customer spending data can also be frustrating. Typically, customers are not as brand loyal as most companies believe (or hope). However, customers are strongly influenced by service, store layout, price, and location. Therefore, in some ways success is easy to quantify using controlled factors—such as location, facility, price, and service—all of which can be displayed visually and geographically. Unfortunately, markets are also beset by uncontrolled and constantly shifting factors such as competition and other changes that disrupt the equation. Ultimately, the more information a business can gather, the better. The highest quality information often comes from actual customer expenditures.

To understand the importance of consumer spending data, refer to the following thematic map, which shows consumer spending on pets. Four categories of spending are depicted: pet food, pet supplies, pet services, and veterinary.

Marketing and Advertising

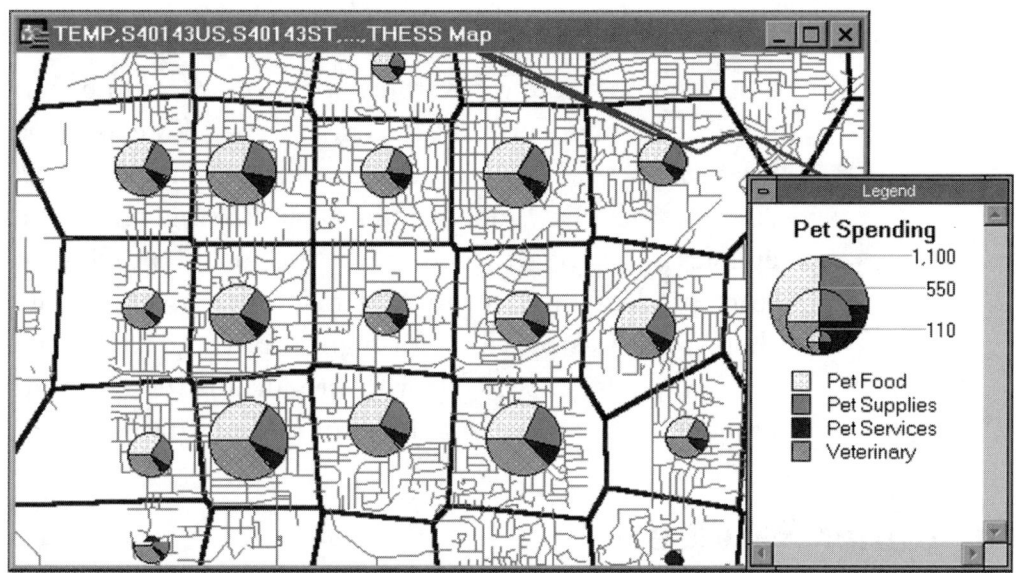

Thematic comparison of consumer spending on pets.

The map illustrates that funds spent on pets varies with geography. While the percentage of dollars spent in each category is relatively constant, the total amount is larger in some areas than in others. A business with limited marketing dollars would certainly want to target consumers who typically spend more on related services.

Consumer Expenditure Survey

Data vendors evaluate spending patterns found in the Consumer Expenditure Survey (conducted by the Bureau of Labor Statistics in the U.S. Department of Labor) and prepare the data in a variety of useful formats. The survey includes spending figures on more than 600 detailed categories of goods and services. Data on spending can be used to estimate potential retail sales for every market in the United States, and some vendors create five-year projections to estimate future spending—information useful for site analysis, product mix, and direct marketing.

Convenience store (or c-store) operators also can benefit from consumer expenditure data. On a store by store basis, operators can examine expenditure data on a variety of products. Understanding purchasing tendencies enables marketers to determine the fastest moving in-store product mix in order to maximize inventory and shelf space. C-stores look for high turnover products, and matching customer needs with products is the first critical step toward increasing revenue and profitability.

Customer Service

Once a business has implemented mapping techniques for marketing and advertising, it can maximize customer service and convenience to maintain a competitive advantage. As such, MapInfo can be an effective strategic marketing tool to help meet the needs of a geographically diverse customer base.

A key element of successful customer service is to ensure that customers can locate your business and that you can locate your customers. MapInfo's locational functionality can play a major role in helping service providers and product marketers satisfy customer needs and expectations. Locating services falls into two categories: locating a customer (e.g., for emergency medical or roadside service, taxi service, fast food delivery, and utility repair) and locating a facility or product (e.g., health club, ATM, post office, gas station, and public attractions).

Customer Service

For example, assume a consumer planning a trip to Washington, D.C., wants to learn the location of hospitals in her health network to schedule a medical procedure during her visit. By calling the health network's toll-free number, the consumer can learn the location and relative distance to participating hospitals. Mapping these data might look something like the next illustration.

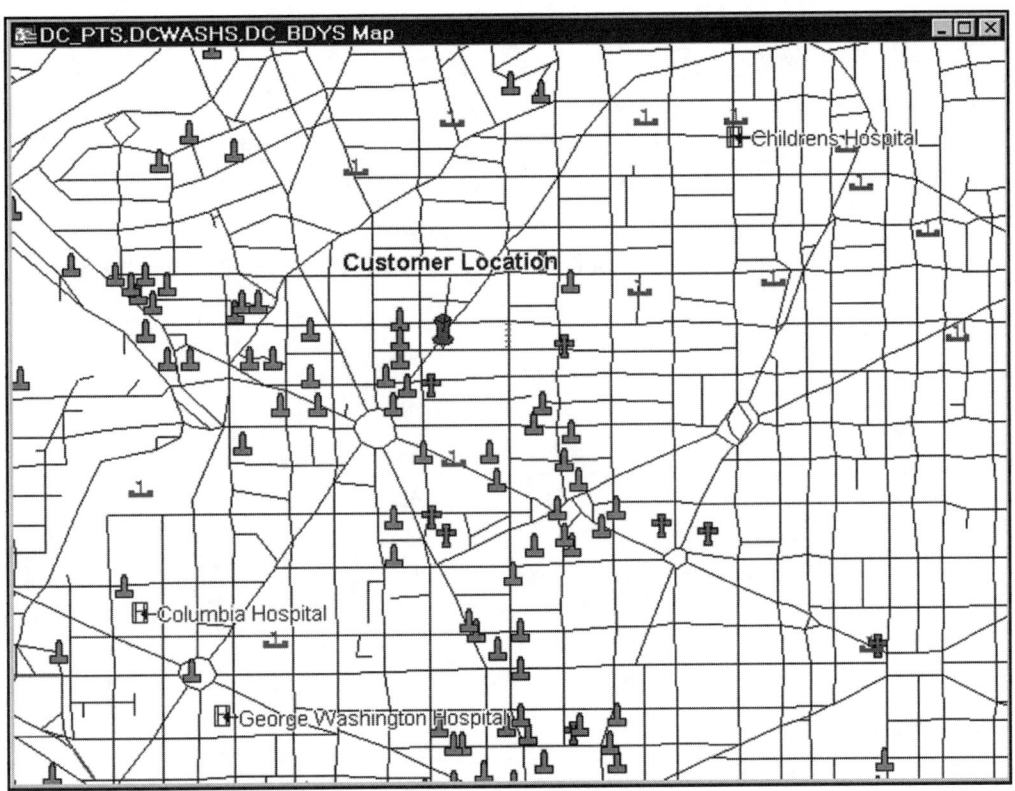

Consumer's hotel (pushpin symbol) and hospitals in the health network.

In addition, the map can be printed and mailed to the consumer if necessary.

Many businesses are using MapInfo's "find nearest" tools to improve customer service via the Internet and toll-free numbers. Retail chains provide this service so that customers can find the location nearest to their home or business. "Find nearest" applications use ZIP Code boundary data or geocoding engines such as MapMarker to pinpoint a customer's location. MapInfo generates a map detailing the customer's location, the location of the nearest retail outlet, and pertinent information such as address, phone number, hours of operation, and distance from the customer.

One of the most prominent examples of retail locators accessed via the Internet is Wal-Mart. The company employs the MapInfo application MapXsite to enable customers accessing the retailer's Web site to locate the most convenient store.

Customer Service

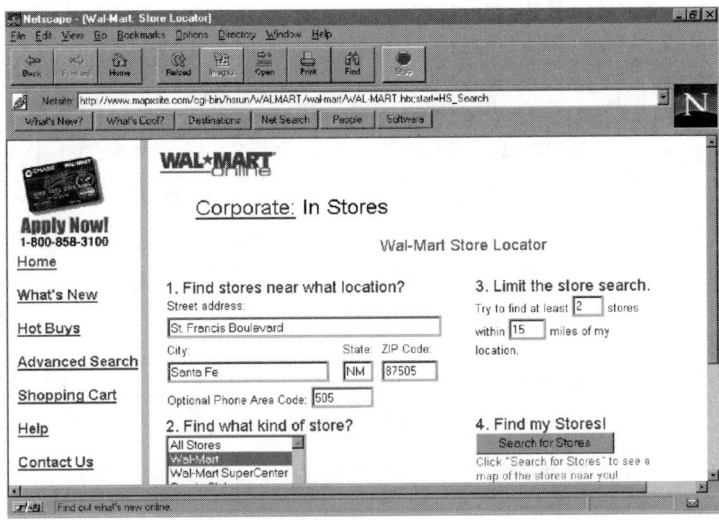

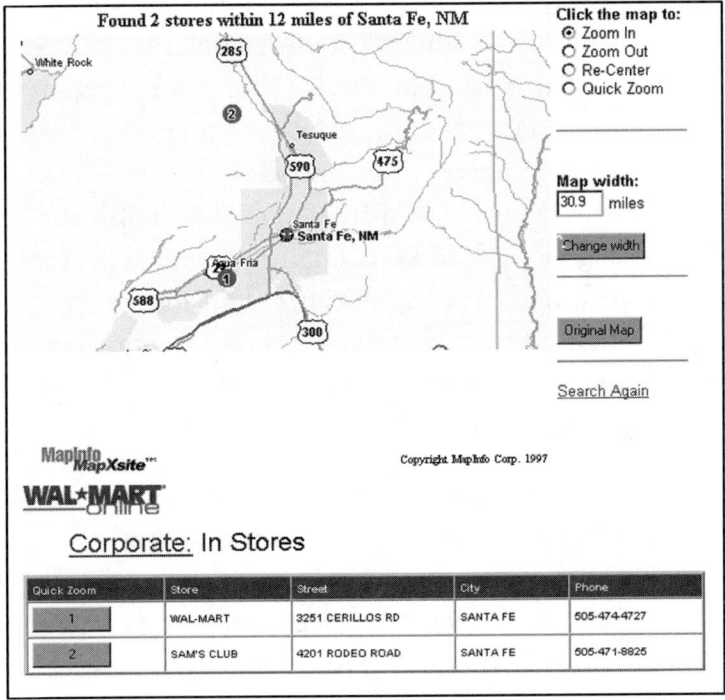

Wal-Mart Store Locator.

Thinking Ahead

In one sense, applications of desktop mapping are limited only by your imagination—and your budget, of course. For example, imagine being able to use mapping techniques and geocoding in order to help a car rental customer who has broken down in an unfamiliar setting. While a hypothetical scenario, developing such a system would require little effort.

When the customer calls a toll-free roadside assistance number, the rental company agent would locate the customer using cross streets, a known address, or even the telephone number of a pay phone. (Of course, if the customer rents a luxury car, it could potentially be located through its cellular phone signal or even via a global positioning system device but such scenarios—in use now—could occupy another chapter in themselves.) Then the agent would geocode the customer's location and generate a MapInfo map displaying the customer's location and service facilities (e.g., car rental offices, repair and towing stations, locksmiths, and so forth). Depending on the nature of the emergency, the agent could also directly connect with the closest emergency response personnel.

Such a system would reduce customer dissatisfaction and cut repair costs for the rental company, particularly because car repair costs are directly related to how far towing and repair services must travel. Using desktop mapping benefits everyone involved.

Site Selection and Analysis

Selecting a business site is becoming increasingly involved and complex. Historically, a general business sense and a review of relevant real estate, traffic, and residential patterns were sufficient to select potential sites. At present, site selection is both more costly and problematic because most ideal corner locations are either already developed or extremely expensive.

Site location decisions are among the most important a business will make. While decisions for small businesses may be determined mostly by their budgets, large companies often employ specialists who select sites using advanced methods.

The discussion of site selection frequently focuses on retail businesses. However, site selection can also be important in locating warehouses, manufacturing facilities, office space, service providers, and even new homes. In each case, desktop mapping tools such as MapInfo enable users to view site locations and surrounding areas, adding a valuable dimension to the process. (Home searches are discussed in Chapter 4, "Banking, Insurance, and Real Estate.")

The following sections focus on desktop mapping applications for retail and non-retail site selection. A discussion of modeling follows.

Demographics

As in marketing, demographic data are integral to evaluating potential sites. In the United States, demographic statistics are usually derived from U.S. Census Bureau data. However, because census figures can be up to nine years old, modeling techniques are often used to project or update census statistics to reflect current demographics. For example, to estimate the population of each census tract within a selected area, a data provider may look at 1980 and 1990 census data and birth and death records to estimate 1997 population.

Snake Graph

A simple review of demographic data such as population or income may be sufficient for some site selections. However, more complex analyses are certainly possible, given the volume of demographic data available. The following "snake" graph, a MapInfo add-on, shows one effective method for reviewing multiple variables.

In practice, the graph enables businesses to plot user-defined demographics for one or more locations against the same variables for competitor locations and/or a market "average." The user supplies selected demographic data (identified in the "Category" column) for all sites to be plotted and the overall market. The "market"—also user defined—could be telephone area codes, ZIP Codes, census block groups, or similar variables. For example, one computer superstore could use the graph to evaluate its performance against specific competitor stores, the market average for all competitor stores in a given ZIP Code, or an industry wide average. Such a technique easily points out the best site based on demographic criteria.

Site Selection and Analysis

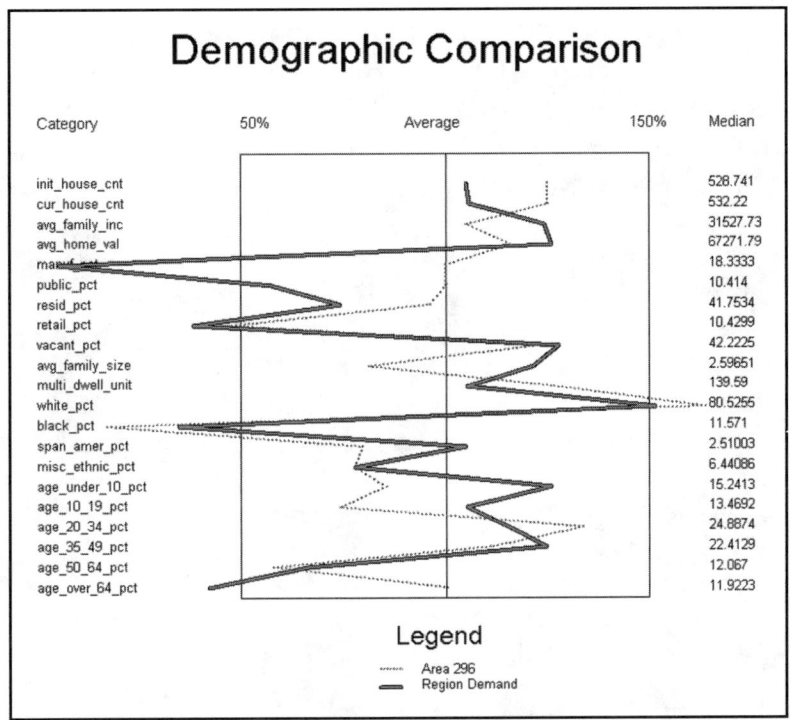

Comparing multiple demographic attributes of a target area (dotted line) and surrounding region (solid line) against a market average (center line).

Five-year demographic projection data enable you to forecast growth and decline according to a specific variable. The thematic map below shows the five-year population trend for one area. Light areas show projected declines, while darker areas show projected increases.

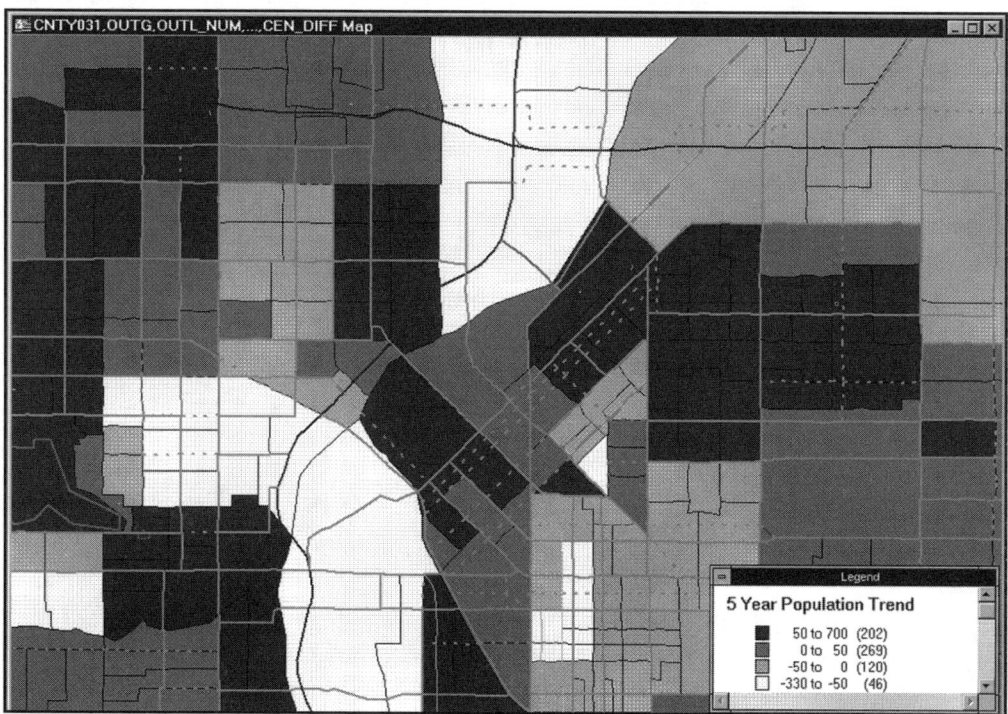

Thematic map showing five-year population demographic trend for a given area.

Thematic maps projecting change can be generated for every demographic attribute. Such an analysis can show anticipated increases and decreases in population and income, as well as shifts in age and ethnicity. Reviewing a five-year demographic projection enables businesses to plan for the future.

Demographic data and their subsequent analysis constitute the cornerstone of site selection. From simplistic views of total population to more complex, analytical and mathematical models, demographics are the common denominator. Demographics can be used to predict sales volumes and profile prospective customers.

Demographics come in all shapes and colors—literally and figuratively. Fundamental demographic data include population counts, household counts, income levels, ethnicity percentages, and age categories. Even such basic information can provide substantial insight. Demographics can be provided at a wide variety of levels: block group, census tract, ZIP Code, county, metropolitan statistical area (MSA), state, and even country. On the other hand, demographic data can be provided in very specific, minute detail, such as the number of Hispanics ages 25 to 34 with an income between $55,000 to $75,000 who drive 10 to 15 minutes to work. For site selection purposes, it is best to have the demographic data broken down to the smallest geographic area available, typically block group. The following narrative describes how demographic data can assist in site selection.

Backing into a Location by Profiling Success

One entrepreneur decided to initiate a site selection for a new health club/tanning facility. With no knowledge of the issues involved, the investor was simply attracted to the idea after observing an existing, extremely successful club. Because the GIS developer had little direct knowledge of the business, the firm chose to "back in" to a customer profile to accomplish the site selection study.

Following the assumption that the existing facility was successful, a demographic analysis of trade areas stretching 1, 2, and 3 mi from the facility was conducted using MapInfo Professional, demographic data categorized according to block group and traffic counts. The data were broken down by age, income, housing status, street traffic flow counts, and total population, and the analysis revealed a profile of the target customer base for the new health club. Based on the profile, the entire MSA was queried to locate several areas that encompassed similar demographic traits. The MSA search area was refined further, and then each potential new site was inspected. Using MapInfo and data, the developer was able to expedite the analysis, systematically refine and define potential locations and provide a solid business plan filled with colorful and informative maps detailing the logic and analytical methods used in the process.

Competition

Competition is also a major factor affecting site selection, whether businesses want to locate far from competitors or across the street. In either situation the first step is to identify competitors and geocode their locations. Such data can be obtained from electronic yellow page listings or data vendors. At a minimum, business listings provide competitor names and addresses. If latitude and longitude information is unavailable, the list can be geocoded so that the data can be viewed on a MapInfo map. Data vendors may also sell competitor data organized according to federal standard industry classification (SIC) codes. Additional data may include store size, number of employees, and sales. The following map shows all competitors for a hypothetical convenience store in a given area.

Site Selection and Analysis

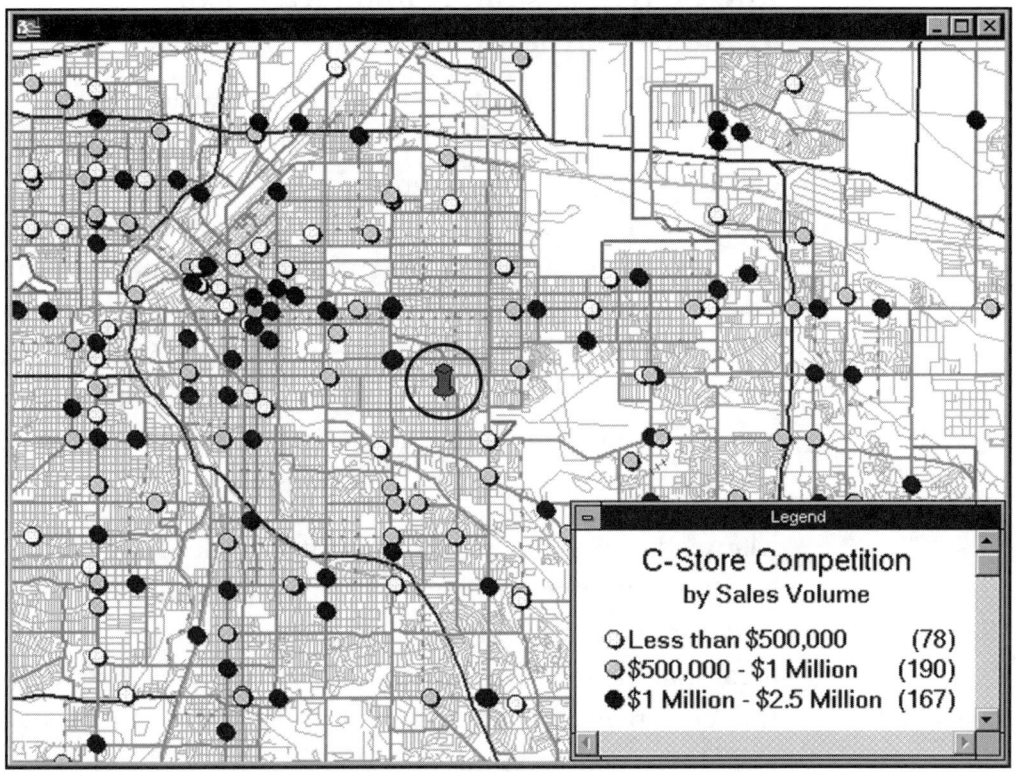

C-store competing businesses and estimated sales.

The map shows relative sales for c-store competitors (darker dots indicate higher sales), and the pushpin near the center of the map represents a potential new c-store site. The map indicates that the concentration of c-store competitors is not high in this area, which may indicate a sound choice. However, traffic flow and accessibility must also be considered. To evaluate these variables, businesses must perform a trade area analysis of potential sites.

Ring, Drive Distance, and Drive Time Trade Area Analysis

A trade area is traditionally defined as the region (or polygon) containing 70 to 80 percent of a site's customers. Three types of trade area analysis are generally performed in retail site selection: ring, drive time, and drive distance. How each is used depends on the nature of the business. In ring analysis, MapInfo draws a series of concentric rings around the site (typically 1, 3, and 5 mi in diameter). Reports are generated to compare inner and outer ring characteristics according to population, age, ethnicity, business activity, and housing.

Many add-on packages for MapInfo are designed to create drive distance polygons, which are particularly important for locating convenience-driven businesses. Trade area analysis may also touch on daytime versus residential customers, which would involve employment and residential demographics.

Site Selection and Analysis

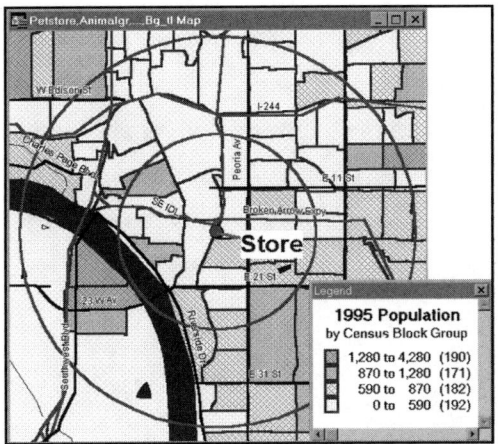
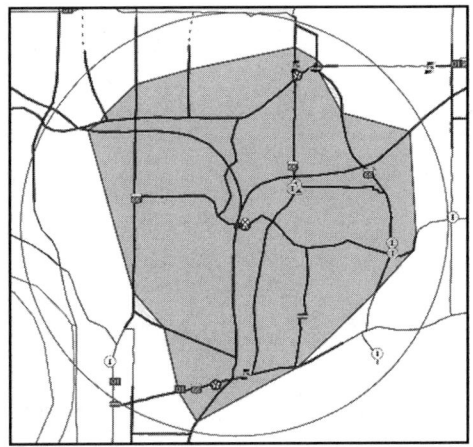

MapInfo tools can determine distance from a given point in three ways: using physical distance or rings (top left), drive distance (top right), and drive time (bottom). Each method can yield significantly different results.

Modeling

The sophisticated technique of modeling enables analysts to calculate the effect a business will have on market share, whether in the context of a new site, changing product mix, or simply improving a site. In general, two site analysis models exist: screening models that "score" site quality, and predictive models that can forecast current or future sales performance. Both types identify key drivers related to performance and provide a method to prioritize new sites and benchmark the performance of existing units.

In order to produce reliable results, you must have accurate, concrete data and know what modeling methodology works best in a given situation. Some models cost more to implement, and the accuracy of results can vary. In many cases, several models are combined to produce a satisfactory result.

Determining the appropriate variables to model for a site's primary and secondary trade areas involves collecting many variables (e.g., business and residential demographics, competition, consumer spending, location, and so forth) and then analyzing their usefulness. Demographic variables tend to influence planned destinations, while a site's physical characteristics tend to affect impulse or convenience purchases. However, even sophisticated modeling and desktop mapping techniques cannot guarantee success. While businesses yearn for perfect site evaluation and selection, they focus most mapping efforts on risk estimation, or they use modeling methods to avoid costly mistakes.

Demographics can be thought of as the DNA, or building blocks, used to model potential demand for a new site. Geographically identifying population counts and other characteristics allows modelers to estimate demand potential at dispersion points within a marketplace. Simply put, this means modelers know where the people are, and they can predict the potential demand that customers generate for a particular product. Once modelers have collected demand information, they can allocate potential demand to existing retail locations in the marketplace in order to replicate the current business environment. By continually adjusting key variables and reallocating demand, the model ultimately accurately predicts sales volume for all existing locations in the market. Once the volume matching process has been achieved for all sites, the market is considered modeled.

Through key variables, the finished model mathematically represents current consumer purchasing habits. Consequently, a modeler can predict purchasing habits in the future marketplace. Market changes (e.g., new sites, closed sites, new hours, prices, and so forth) can be modeled to calculate new sales volume figures. The ability to run multiple scenarios on different locations and predict the sales outcome before investing money is a very effective planning tool. Marketers can also model potential competitor behavior to determine how it may affect company sites.

Many firms provide modeling services in various formats and technologies. Undoubtedly, all rely on basic demographic data to build these incredible tools. For another discussion of demand modeling, see the "Canada Post" case study in Part Two.

Sales Territory Redistricting

The process of assigning and reassigning map objects to groups or districts, or redistricting, is a valuable MapInfo function often employed to realign sales territories, school and voter districts, emergency service coverage areas, delivery routes, natural resource management areas, and electronic communication coverage areas, among other uses. The following images illustrate MapInfo's redistricting function.

Sales Territory Redistricting

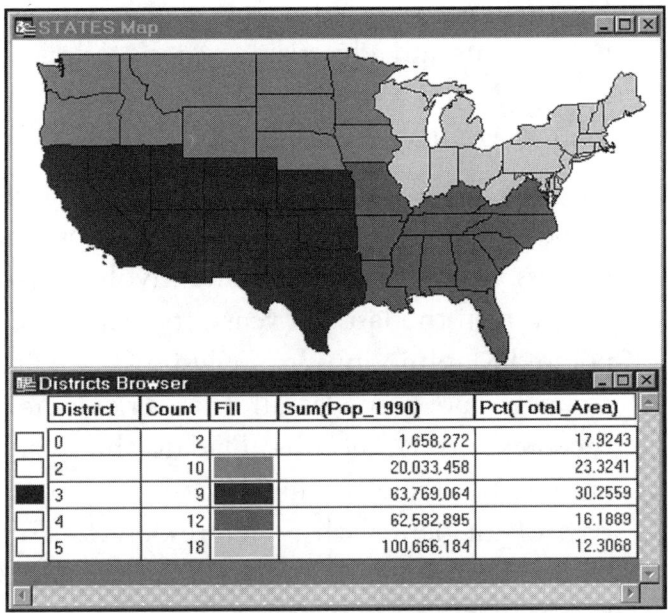

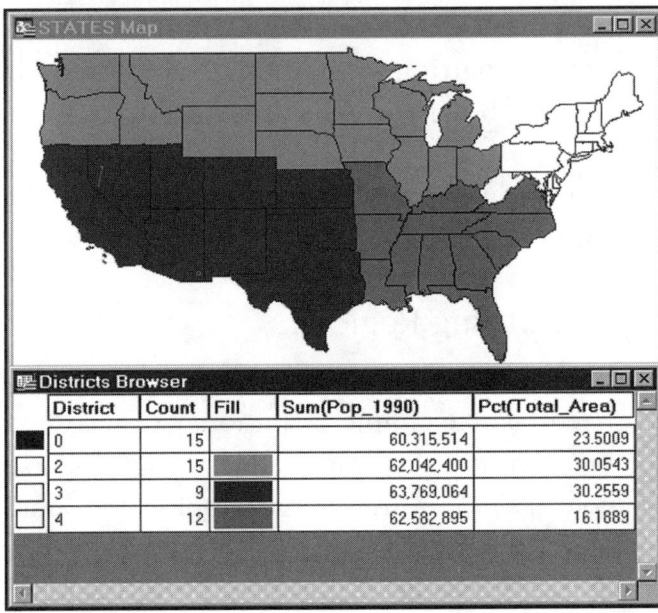

Reallocating areas from District 5 (top) to Districts 0 and 2 (bottom) is accomplished quickly in MapInfo, making it possible to rapidly test multiple scenarios.

Discussions of redistricting successes with regard to police beats and advertising regions follow.

Police Beats

City X's 911 communications department must redistrict its police beats annually. Using traditional methods, the process took several weeks and involved numerous officers. Within the last few years, the department has incorporated MapInfo into its analysis. The redistricting team captures and geocodes all crime incidents occurring in the previous 12 months. Placing the events on a map enables the team to quickly compare crime density and type of incident with police resources. The maps effectively indicate gaps in coverage that could result in a higher crime rate. As a result, the team is able to logically allocate more resources in high crime areas—in essence putting cops in high risk areas before crimes occur. The process has been an effective crime deterrent because it improves resource allocation by requiring significantly fewer officers to complete, and reducing the time involved.

Advertising Regions

MapInfo's redistricting tool can also support advertising needs. For example, a national service company had a method of allocating advertising dollars around the country according to geographical boundaries. While geography was an effective starting point for the process, it failed to take into account the volume of customers and locations within each region, which unfairly compared regions against one another. The company, which had implemented MapInfo for other purposes, subsequently discovered that it could also use the software to realign its advertising needs to better reflect customer volume and retail locations. Still using geography as the driving force, the company adjusted its advertising region boundaries to equalize customer and location figures in each region. As a result, regional managers stopped competing in unproductive ways and funds were fairly allocated to each region.

2

DECISION SCIENCES

In general terms, the discipline of decision sciences combines expertise in organizational and behavioral sciences with statistical research methodologies and computer science. Desktop mapping applications in this area involve making decisions that affect the movement of people and goods taking into account numerous variables. While MapInfo solutions in the realm of decision sciences are just beginning to be explored, a number of exciting possibilities have emerged. These include inventory management, transportation applications, and routing.

Inventory Management

Prior to the Great Depression, inventory management was relatively straightforward for many businesses. In fact, retailers, producers, and policymakers often counted inventory as a measure of wealth. This view changed dramatically during the Depression, as a rapid rate of inventory turnover became the goal. Today, inventory management strives for balance. Large inventories are viewed as risky rather than a measure of wealth, and excessive merchandise may require drastic sale pricing if its worth declines. This section covers how MapInfo can be used to help visualize and analyze inventories in retail and warehouse environments.

Successful retail stores are characterized by a tremendous amount of inventory turnover. The volume of products that move through popular supermarkets and convenience stores on a daily basis is quite large. MapInfo can help both retailers and distributors to maximize inventory turnover.

Beverage Company Deliveries

In the extremely competitive soft drink market, product placement at convenience stores can be critical to a local promotional campaign's success. Beverage company drivers must have accurate information regarding where promotional drink displays should be placed. However, confusion often exists between delivery drivers and convenience store managers regarding appropriate product placement. Beverage company X initially sought a CAD (computer aided design) solution to the problem, believing that detailed illustrations—which even included the position of electrical outlets—would eliminate confusion. Approaching the problem from this angle cost the company substantial time and money, and it failed to improve communication, which was the heart of the problem.

In contrast, a MapInfo developer relied on the software's integration and rapid application development tools to quickly generate a prototype solution that focused on communication. First, the developer started with a simple store layout, including the lot, building, and gas pumps.

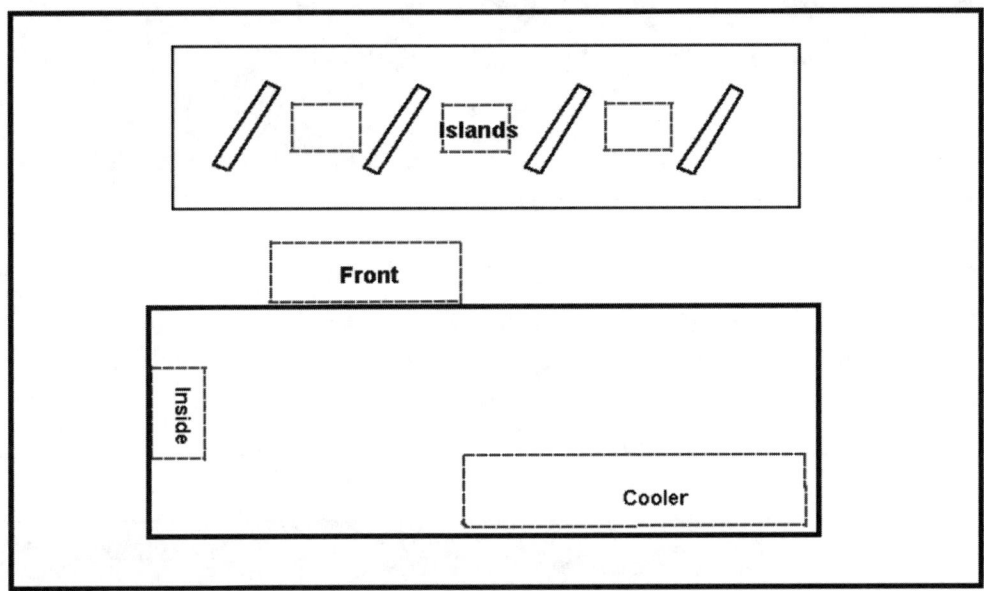

Convenience store layout.

The diagram can help ensure that a delivery truck driver correctly places displays because it can indicate locations (e.g., the pump island and inside at the cooler and fountains) where products should be placed. For further clarity, the driver can click on the appropriate spot to view a photograph of the correct placement.

Pump island (left) and cooler (right) selected from the convenience store map.

Of course, MapInfo can also manage product data. By clicking on the map or selecting a menu item, the delivery person can quickly determine how much of a given product should be placed at a particular location. The following table illustrates a delivery report.

SKU #	Product Description	Size	Quantity
4328907432	Cola	6-pack	10 cases
4325443243	Cola	2-Liter	24
4325456455	Orange	20 Ounce	2 cases

Product delivery information accompanying the map.

Finally, MapInfo's analytical features can help soft drink companies target products to the correct markets. For a more detailed discussion of this topic, see Chapter 1, "Marketing, Advertising, and Sales."

Supermarket Product Placement and Inventory

When placing products in a grocery store, beverage companies have additional factors to consider. If the product appeals to working mothers with little time to shop, the product would ideally be displayed near the express checkout lane. If the product appeals to teenagers, a vending machine near the store's entrance would draw far more business than one inside the store. In either case, product placement could be easily and accurately expressed on a map.

Moreover, in today's market, grocery store size and product offerings determine store survival. Supermarket managers can use MapInfo in several ways to help ensure success. At present, many new grocery stores are smaller than their predecessors because of the sheer cost required to operate large stores, especially when 45 to 50 percent of the inventory is perishable. New stores built today are typically 50,000 to 70,000 sq ft, while stores constructed a generation earlier are 75,000 to 90,000 sq ft.

MapInfo can help site new stores successfully and target marketing efforts to appropriate customers. In addition, analyzing business geographics can help store managers adjust inventory levels to reflect the type of customers in specific trade areas. Existing stores can track scanner data captured during each purchase to gain valuable insight into patron buying habits.

MapInfo can also be used to manage inventory. Of course, maps of shelf inventory must be far more detailed than convenience store maps. The following images show sample layouts of a grocery store produce section.

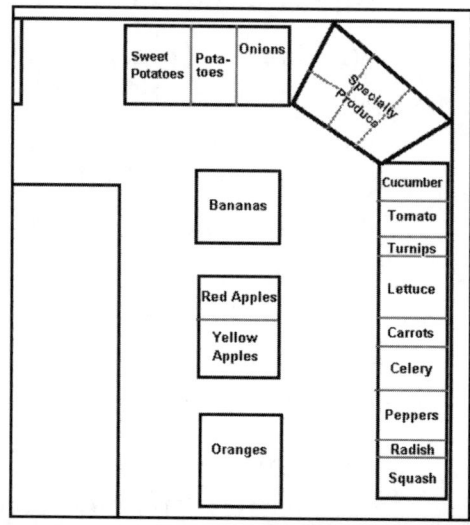

Produce section map (top) and photograph (bottom).

Store managers can analyze scanner data to reevaluate the layout of the produce section and rearrange it if necessary to maximize profit, factoring in produce cost and availability. The map can also be used to schedule checks of the display, stock replenishment, and even employee hours.

Inventory Management

Parking Lot Inventory

While it may be frustrating to misplace a car in a parking lot, rental car companies and automobile importers encounter such situations on a daily basis. MapInfo can offer effective solutions to this problem as well. Car handling companies can use GIS techniques to map lots in order to maximize space and retrieval of vehicles. By mapping vehicle locations as they are processed, companies are unlikely to misplace cars, thereby increasing profitability.

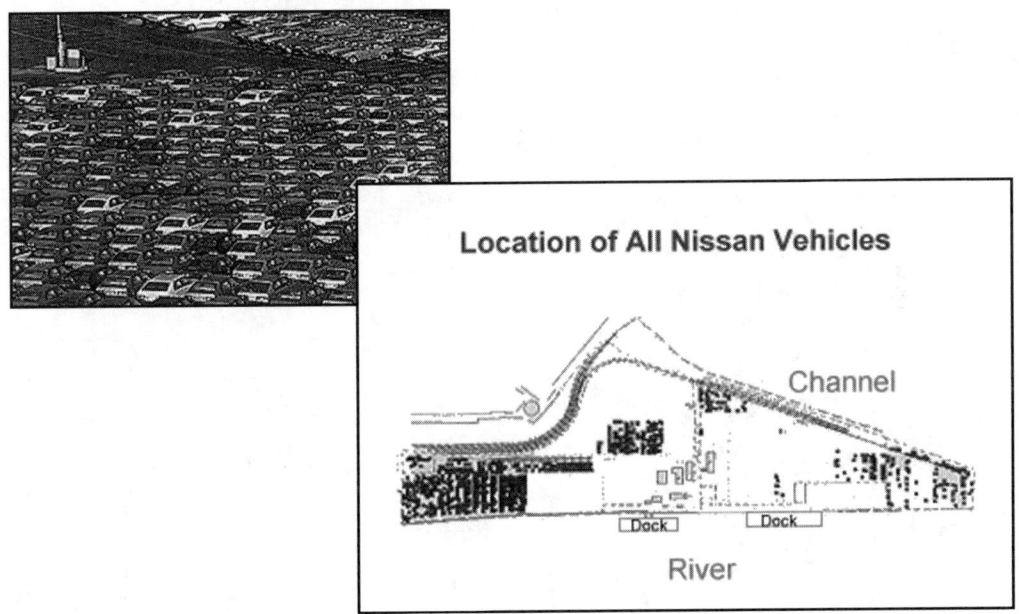

Parking lot (top left) and import car terminal map (bottom right).

Warehouse Inventory

Inventory management is central to many types of entities, not just retail stores. Financial and insurance companies manage repossessed items, large furniture chains manage inventory to prevent excess movement of pieces, and government and military entities routinely track inventory stored throughout the country.

Desktop GIS capabilities in this area are robust. With the proper software, you can use MapInfo to issue a query regarding the location of an inventory item, click on the image of a location where the item is stored, and view a location layout. The following illustrations depict this process.

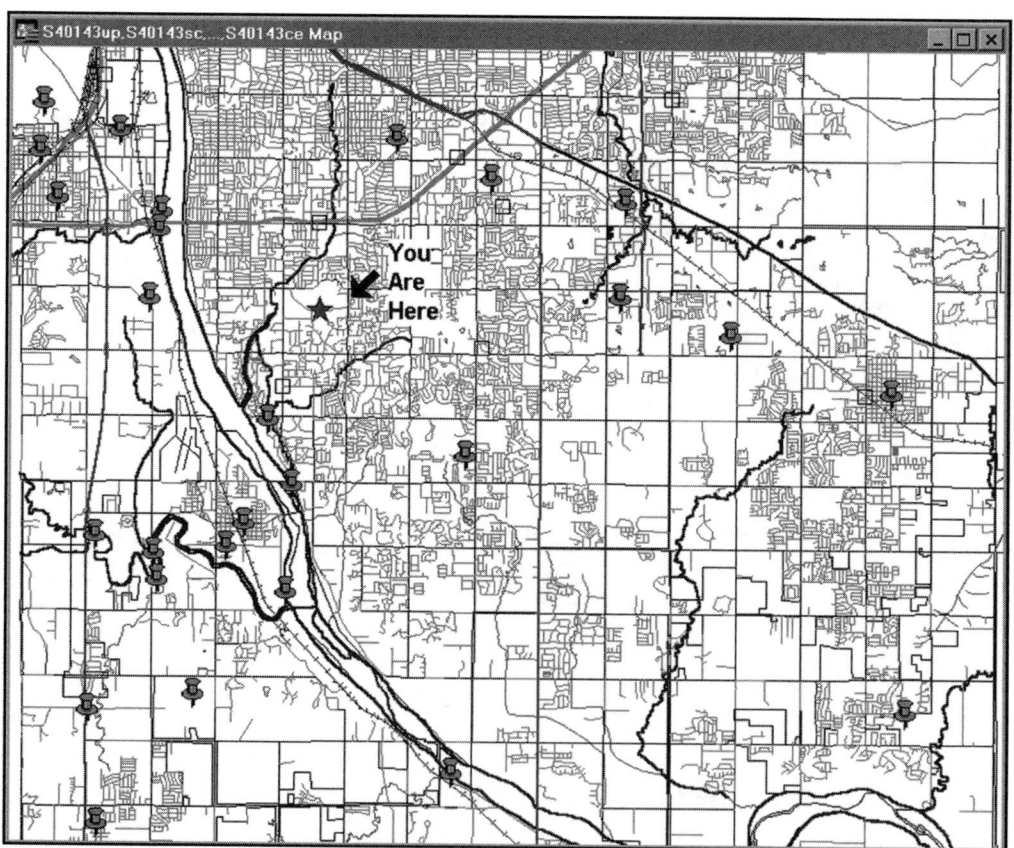

Facility map showing where inventory items are stored (pushpins) compared with the user's location (star).

Inventory Management

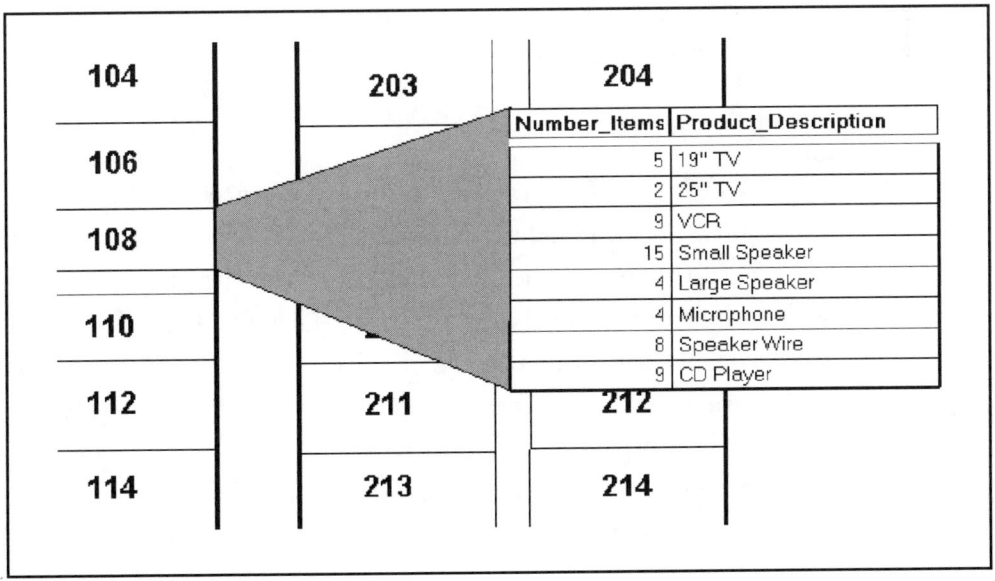

Building layout.

Inventory contained in a room.

Keep in mind that MapInfo functionality can accommodate more complex needs. For example, furniture companies could supplement the previous visual information with detailed illustrations indicating inventory stored along vast warehouse aisles. Furthermore, MapInfo can also help you locate the fastest route to collect or deliver inventory. This topic is discussed in the "Routing and Logistics" section that follows.

Transportation

Desktop mapping applications in the transportation industry are diverse. MapInfo software is used in air, rail, water, and road transportation systems to help planners evaluate network models, reroute traffic more effectively, and schedule maintenance to minimize impact on the public, among other applications. How the railroad industry uses GIS offers insight into the breadth of potential applications.

Railroads

More than 600 rail companies operate in the United States, Canada, and Mexico, managing lines that span 147,000 mi. To help them navigate this vast network, rail companies rely on accurate data and GIS. Major rail data are found in the U.S Railroad Database, which contains line data representing all major and short-line railroad companies. The line data (graphic objects) are linked to a database that contains the following line specific information.

- **Rail operators.** Rail operators usually own their lines, although in some cases lines are owned by one entity and operated by another. The database table contains up to three operator fields to identify railroad companies operating specific lines (the first field is the primary operator).

- **Track rights.** Track rights indicate that railroad companies have agreements with line owners to operate their trains over specific lines. Up to three fields are used to identify companies that have track rights over a particular stretch of rail.

- **Passenger trains.** If applicable, a field identifies passenger service operating over the rail line.

The U.S. Railroad Major Systems Database represents the railroad network for major railroad operating companies such as Burlington Northern Santa Fe (BNSF), Northfolk Southern (NS), Southern Pacific (SP), Union Pacific (UP), and Wisconsin Central (WC).

Real Life Rail Mapping

A major railroad company looked to GIS for assistance in complying with a Pipeline Safety Advisory Bulletin (also known as the Rail-Pipeline Emergency Plans Coordination). The federal bulletin mandates that the location of underground pipelines carrying hazardous material and natural gas must be identified and incorporated into emergency response planning and procedures.

To help comply with the bulletin, the railroad needed information on all pipelines that share, run parallel to, or cross its right-of-way. A GIS consulting firm provided research, database development, and mapping services to create a digital map database of pipeline crossings. Mapping software and training were provided to ensure continued access to and maintenance of the database.

All pipelines in the proximity of the railroad's entire system were identified. In the database, pipelines were coded with the following attributes: pipeline company name and emergency contact, commodity (e.g., natural gas, crude oil, or other petroleum products), and diameter. Key railroad data, including mile post and operating division information, were also provided.

The railroad received a complete database, system maps, state maps, software, and training. In the event of a rail incident, the company can now refer to the database and maps, and if a pipeline is present, coordinate emergency response planning with the pipeline company.

For more information about GIS and railroad data, contact DeskMap Systems (see Appendix D for contact information).

Transportation

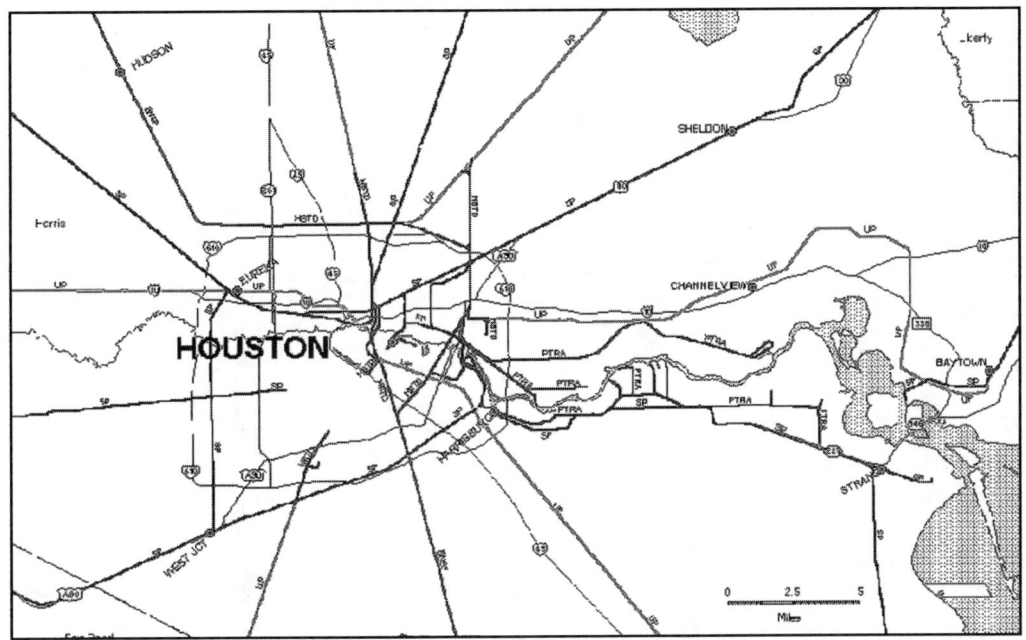

Houston area railroad system.

Railroad data are used extensively for various strategic planning activities such as merger/acquisition analysis, decision support analysis, and flow of goods and commodities.

Rail companies must also analyze profitability. What types and quantities of goods create the most profit for the railroad company? The following maps illustrate profitability and revenue by line segment. The first map is thematically shaded to show profitability (lighter lines indicate profitable segments). The second map uses pie thematic charts to analyze the revenue generated by transporting agricultural, automobile, chemical, coal, and construction products.

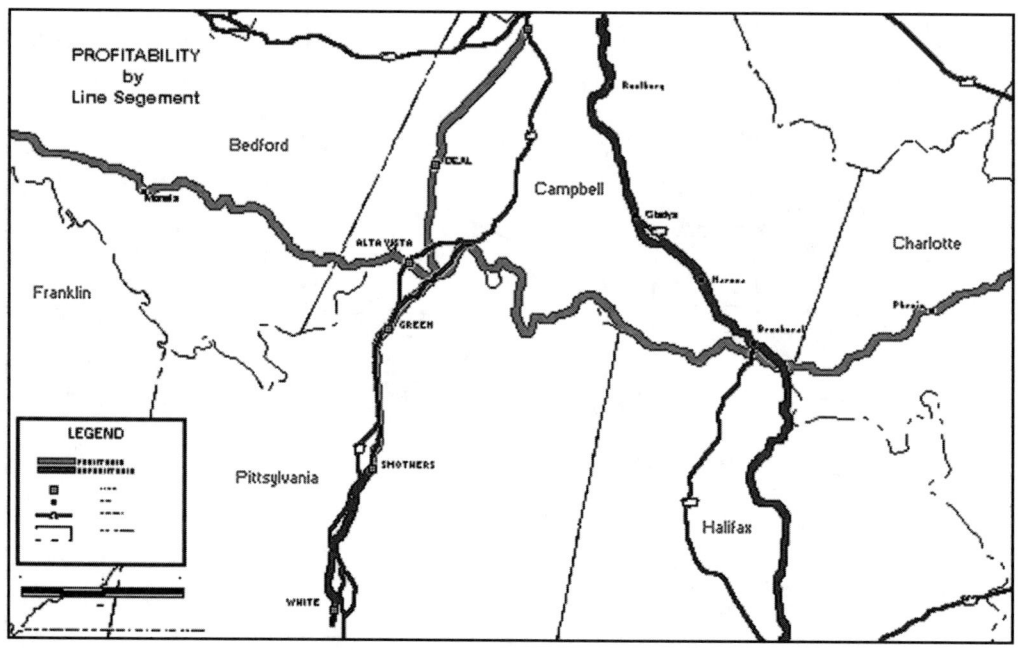

Railroad profitability by line segment.

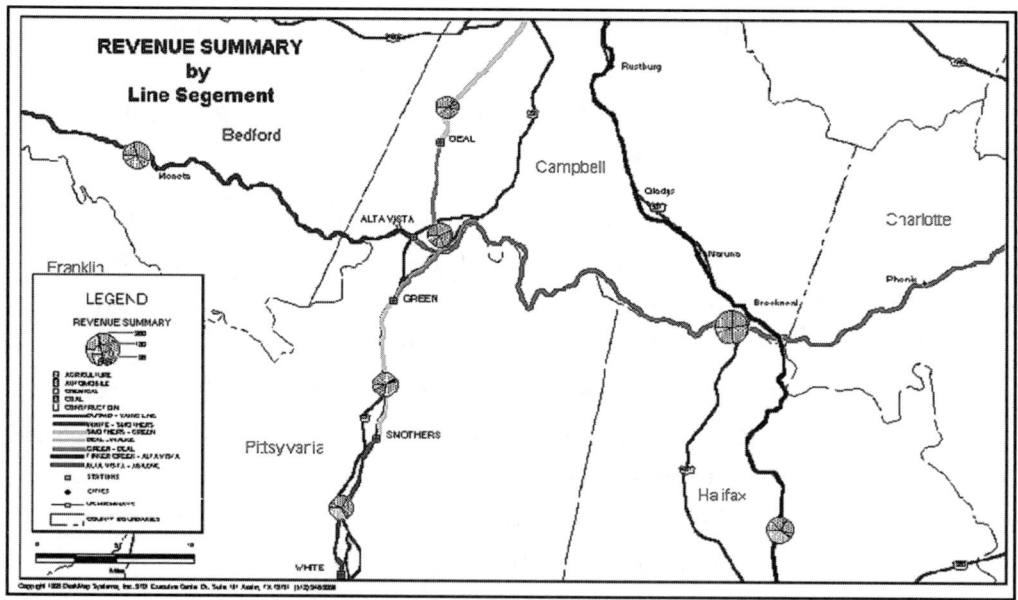

Railroad revenue summary by product.

Businesses throughout the country (and indeed North America) ship a wide variety of commodities by rail. They use MapInfo and railroad databases to track where products are in the shipment process in the same manner that packages are tracked by express delivery companies.

MapInfo is also extremely useful for managing the location of railroad cars. Owned or rented by railroad companies or businesses themselves, the rail cars must be tracked for logistical reasons. Additional information (e.g., distance traveled, destination, and contents) can support profitability studies, maintenance, and long-term planning.

Telecommunications and Emergency Response

With the advance of GIS technology and expanded use of MapInfo, railroad data are finding increased use in other industries, particularly telecommunications and emergency response. Because the railroad infrastructure connects major metropolitan areas, the link between railroads and telecommunications is evolving into an effective partnership. Telecom companies are attracted to railroad right-of-ways for placing new fiber optic cable and additional transmission towers. Emergency response organizations rely on railroad data because railroad cars often carry large quantities of hazardous materials such as fertilizers and chemicals. Knowing where toxic chemicals are located can impact emergency response procedures and evacuations following derailments or accidents.

Intermodal Transportation

The term *intermodal* refers to the combination of different modes of transportation. Certain locations in the United States serve as intermodal transportation facilities, or points where goods are transferred from one method of transport to another. The Alliance Airport in Dallas is one example. Baltimore, another example, combines rail, truck, and water transportation systems.

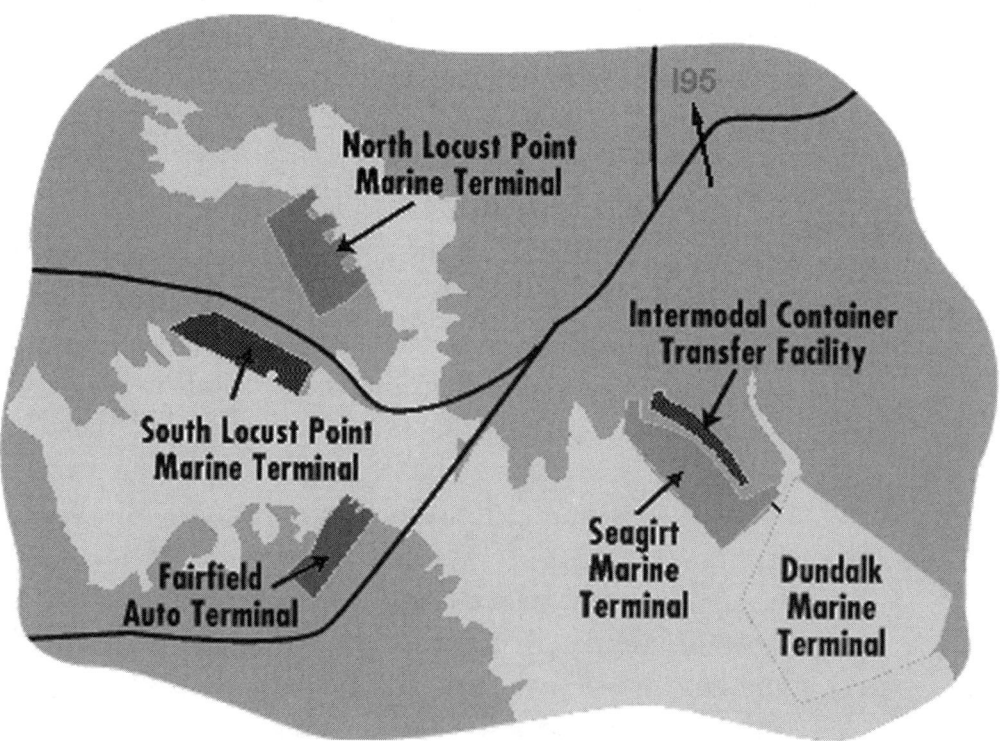

Baltimore intermodal transportation facility.

MapInfo is used in trucking, air freight, and water transportation in much the same way it is applied in the railroad industry. Because of its horizontal nature, MapInfo can be easily applied in intermodal facilities as well.

> ✓ **TIP:** *The Bureau of Transportation Statistics within the U.S. Department of Transportation is a major source for transportation related data. For a list of available data, see Appendix B, "Reference Material and Data Sources."*

Roadways

Transportation professionals responsible for planning, designing, and operating street networks and transit systems use desktop mapping in several important ways.

Planning and Traffic Counts

Transportation planning encompasses the design and construction of new roads, organizing and tracking maintenance, and even researching innovations. MapInfo is used to help planners evaluate potential roadway designs against a variety of parameters, incorporate modeling methods to address population changes, present the results in a visual format that eases public and environmental concerns, and support traffic impact analyses.

Traffic counts are one tool traffic engineers use in planning. Combining modeling techniques with mapping software to display counts can help engineers obtain an inexpensive yet thorough coverage of traffic counts across a city.

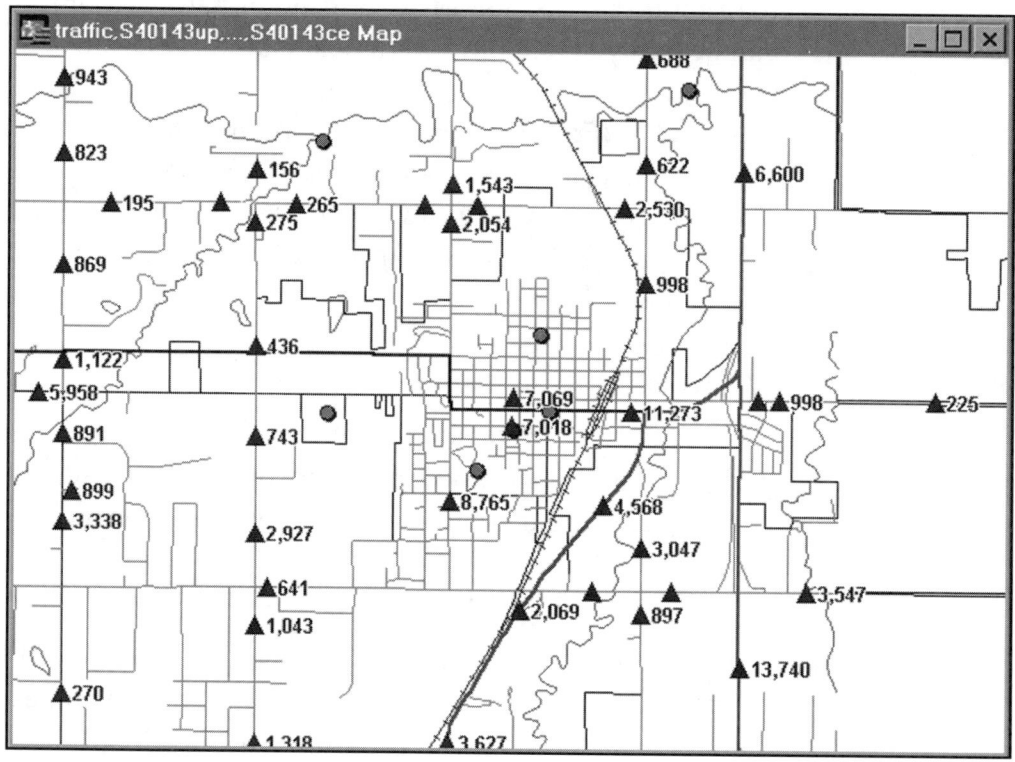

Traffic counts (triangles) for intersections in a given city.

MapInfo's presentation capabilities can also be extremely useful for interacting with elected officials and the public regarding new roadway designs. Environmental and quality of life issues can be addressed quickly, with the potential of seeing new designs approved more rapidly.

Construction and Operations

Desktop mapping applications for design and construction include traffic signal and interconnect design, channelization and signing design, advanced traffic control systems, street lighting design, corridor signal progression analyses, and construction management. In addition, MapInfo can be used to manage parking systems (e.g., parking meters, time zones, and bicycle facilities). Painting road lines is yet another useful application.

Mass Transit

Public transportation professionals are charged with determining routes and service frequency to adequately meet public demand. MapInfo is being used as a planning, management, and information system for public transport service operators.

With MapInfo and custom built add-on tools and menus, transportation professionals can maintain geographical data for networks to assist in planning and scheduling. Where bus stops are positioned, the location of train stations and bus line routes can be edited in a digital map with technical and sociodemographic attributes. Because MapInfo can interface with many SQL databases (e.g., ORACLE), mass transit network systems can be adapted to virtually any other planning and scheduling system with a relational database (RDBS).

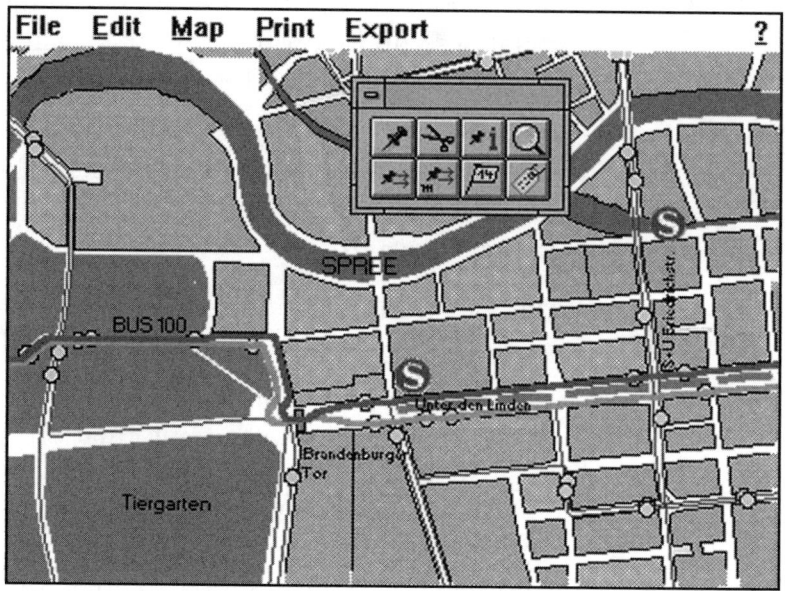

MapInfo custom interface for managing public transit.

MapInfo can also produce a variety of printed maps with route and network information for dispatchers and drivers. In addition, dispatchers can use MapInfo to access the current network data directly. Finally, creating high quality printed schedules and maps is easily accomplished in MapInfo (see the next illustration). Housing data in a digital format facilitates changes and updates to the network.

Transportation

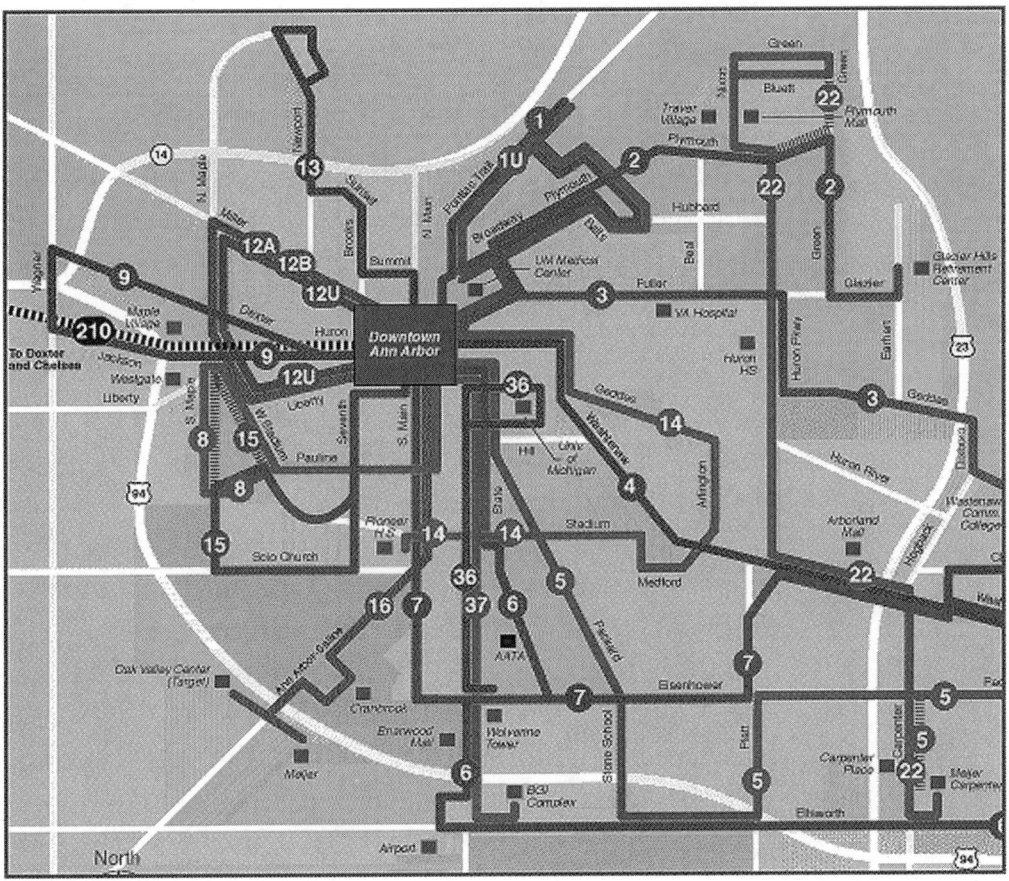

Public transportation map.

Global Positioning Systems

Global positioning system (GPS) technology has matured into a valuable resource. At present, scientists, hunters, farmers, soldiers, pilots, surveyors, hikers, delivery drivers, sailors, dispatchers, loggers, firefighters, and others use GPS to make their work and recreation more productive, safer, and often easier.

GPS allows you to calculate a precise position on the Earth's surface (i.e., latitude and longitude). This technology, developed by the U.S. Department of Defense (DOD) to simplify accurate navigation, is comprised of a system of satellites (NAVSTAR) orbiting at high altitudes and communicating with hand-held receivers on Earth. The exact coordinates of the GPS receiver are calculated by measuring the distance from NAVSTAR satellites to the receiver. Although the federal government has invested billions of dollars in GPS, data from the NAVSTAR system are available at no charge.

Common uses of GPS technology include fleet and emergency vehicle tracking and navigation in rental cars, aircraft, and shipping. Potential future uses of GPS include collision prevention and accurate zero visibility landing systems for airplanes. (Such applications require data transmitters as well.)

Inclement weather and tall buildings can degrade the accuracy of GPS readings. However, the DOD itself is the major source of signal degradation; the DOD deliberately degrades signal quality to obscure the location of military installations and artifacts. Recently, the DOD was ordered to develop a plan for the elimination of such intentional degradation, known as *selective availability*. Another process, known as *differential correction*, can also be used to correct signal inaccuracies. Differential correction compares GPS coordinate readings for known, accurate points with current GPS data; the difference is applied to other GPS data in the survey.

Innovative GPS Solutions

Power Company

A regional power company (with a combined total of 27,000 mi of power lines and approximately 330,000 poles) implemented an AM/FM/GIS system to map the poles, and an independent pole inspection company began implementation of a differential GPS system to collect the inspection data. Pole coordinates are collected in North American Datum (NAD) 27 latitude/longitude and then converted to state plane coordinates for mapping. Now, both sets of data are easy to maintain and visualize in the mapping system.

As with many projects, the pole mapping endeavor began with a trial attempt to map a 6 mi-long line. Far more extensive work has taken place since the initial trial was conducted. The differential base station is set as close as possible to the circuit to be mapped. Baseline distance is kept within 30 mi where possible. Accuracy measured against local benchmarks was found to be within 1m to 3m, more than adequate for the application.

Nuclear Plant

In 1993, the Westinghouse Hanford Company began to operate a tractor-based system that uses GPS, GIS, and radiation detection technologies to survey contaminated areas around the Hanford nuclear plant.

It was common practice in the early years of the nuclear industry to dispose of slightly contaminated waste directly in the soil. Records of such disposals were not always complete. Before the GPS system was implemented, work crews in protective gear would walk the contaminated area. The inspectors carried hand-held instruments and would make notes on hand-drawn maps. The results of these inspections were unclear and imprecise.

After three years of conceptualization and design, Westinghouse Hanford built a specially equipped tractor called the Mobile Surface Contamination Monitor (MSCM-II). The MSCM-II went into operation in May 1993. The tractor is an 18-ton vehicle with a modified front loader that carries three pairs of scintillation radiation detectors. The rear cab contains the radiation detection system, a GPS receiver, a VHF radio, and a 386 PC and monitor. The PC provides real-time data and navigation displays. An operator rides in the rear cab with the equipment, and a driver rides in the front cab, controlling the detector assembly.

The GPS system uses differential techniques to provide the necessary accuracy. Nine channel CA code receivers are used. The GPS in the tractor is linked to a base station by a VHF radio. Differential information is transmitted to the tractor using the RTCM-104

protocol. New position updates are calculated every second and recorded in the PC.

The first area surveyed spanned 165 acres. In 27 days, the MSCM-II was able to survey 100% of the area, collecting 248,437 data points. The survey identified 75 previously unknown areas with elevated radiation levels. Approximately one ton of material identified in the process was discarded. The MSCM then surveyed a 79-acre area in 13 days and collected 285,810 data points. Data quality has dramatically improved, resulting in a $5 million dollar savings, and greatly increased worker safety.

GPS receivers are currently available in navigation, mapping, and survey grades. As accuracy requirements for exact positioning increase, so does the cost of the GPS receiver. Navigation GPS receivers are the least expensive and are accurate to the length of a football field (approximately 100 yd). They do not interface well with PCs and are not equipped with differential correction capabilities. Mapping grade receivers, guaranteed to be accurate up to 1m, are equipped with differential correction capabilities and receive signals from four satellites to calculate position. Survey grade receivers, the most expensive type, are accurate to 1cm. They use more than four satellites to calculate position and acquire sample readings over a long period of time.

Using GPS to precisely survey and map study areas saves time and money. GPS makes it possible for a single surveyor to accomplish in a day what used to take a team of individuals weeks to accomplish. In addition, the process is more accurate than ever before.

Mobiletrack

The Australian company Spectrum Global Telecommunications offers the Mobiletrack GPS solution, which combines advanced signal correction and MapInfo to track vehicles and provide intelligent fleet management and security. In real time, Mobiletrack stores and displays a vehicle's position on a digital MapInfo map that indicates speed and direction. In addition, the system can monitor numerous alarms with the incorporation of other sensors.

Furthermore, GPS technology in Mobiletrack's in-vehicle data terminal enables leading edge fleet management and dispatch solutions that can map and dispatch the closest or most suitable vehicle for a given task.

Mobiletrack Pty Limited developed and uses a real time differential correction (RTDC) system able to continuously compensate for selective availability, generating an industry benchmark of 3m to 5m accuracy. MapInfo was chosen because it represented an inexpensive, turnkey solution that could import a variety of international data sets. In addition, MapInfo's flexibility offers sufficient detail, breadth, and user-controlled functionality. Up to nine maps depicting different vehicles can be displayed at once, which provides easy visual coordination.

Circulation and Delivery

Organizing newspaper circulation can be an enormous task but it becomes manageable with MapInfo. In addition to organizing the delivery process itself, desktop mapping can improve customer service dramatically, enabling newspapers to retain subscribers in an era of declining readership. Plotting all newspaper subscribers in an electronic format offers publishers several advantages. The data can be used to automate, route, and balance carrier delivery. Using the map to help identify a reader who calls with complaints can help pinpoint chronic problem routes or carriers. In addition, the publisher can overlay population information with customer data to pinpoint areas where additional subscribers can be targeted.

> **NOTE:** *Not all circulation and delivery operations require GPS technology. However, most benefit from mapping techniques.*

Newspaper delivery addresses.

On a larger scale, consider the U.S. Postal Service, which must organize deliveries to every mailing address in the country six days a week. Plotting each postal customer on a map is essential for managing mailing locations and carriers. Using GIS technology, the post office can far more effectively manage and balance carrier routes and loads to maximize efficiency.

GPS technology can be used to track vehicles in real time. This enables postal trucks to organize a greater number of pickups and deliveries, which is particularly useful for expanding express mail or other premium—and profitable—services. Now within the reach of many businesses, GPS units can also be equipped with panic buttons to increase driver safety. Such systems can also simplify route analysis.

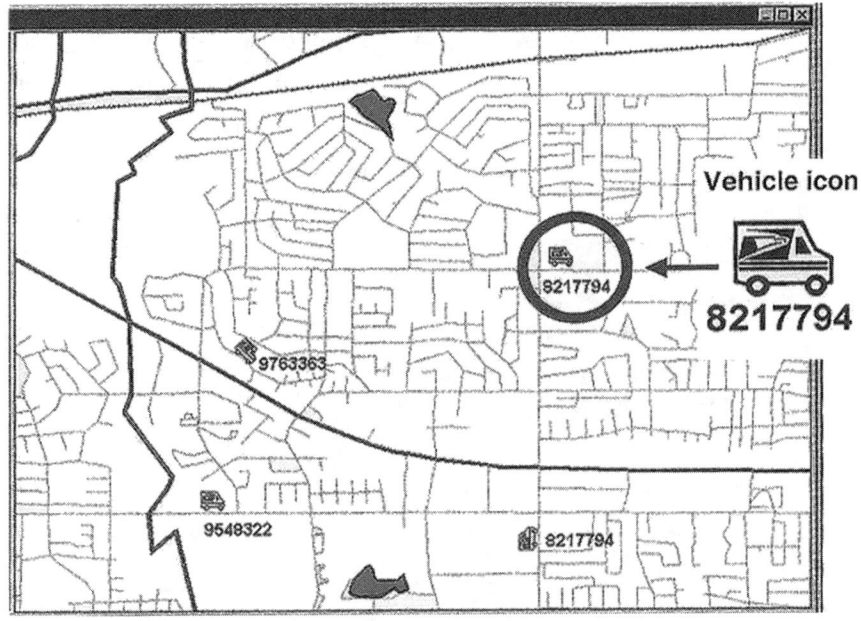

Mapping delivery routes using GPS technology.

Geographic Tracker

The Geographic Tracker by Blue Marble Geographics, Inc. (which ships with MapInfo Professional) includes a MapBasic application (*Geotrack.mbx*) that enables users to perform GPS tracking by showing real time derived positions overlaid on a MapInfo map. One of a collection of GeoObjects, GeoTrack is the OCX (ActiveX) version of Geographic Tracker available for MapX developers. The software also permits users to collect and incorporate field information in real time directly into MapInfo tables. Blue Marble calls the concept "GPS geocoding." You can practice generating GPS simulation files for your locale with the *MakeGPS.mbx* application in Geographic Tracker.

With Geographic Tracker and MapInfo, users have real time mapping technology at their fingertips. Vehicles equipped with GPS receivers can easily be tracked on MapInfo maps in the vehicles themselves.

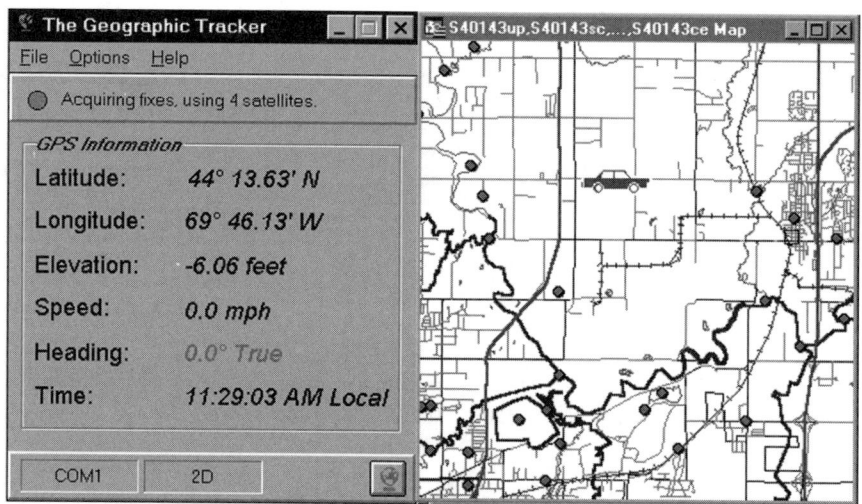

GPS reading of vehicle location in Geographic Tracker.

In vehicles, the technology allows drivers to see their locations relative to their destinations. Using commercial off the shelf (COTS) components such as GeoTrack and MapX in standard programming environments such as Visual Basic, Delphi, and Visual C++, hundreds of delivery vehicles could be routed and tracked efficiently. In a dispatch environment, the technology enables operators to organize service more efficiently.

Geographic Tracker also displays satellite configurations and signal strengths.

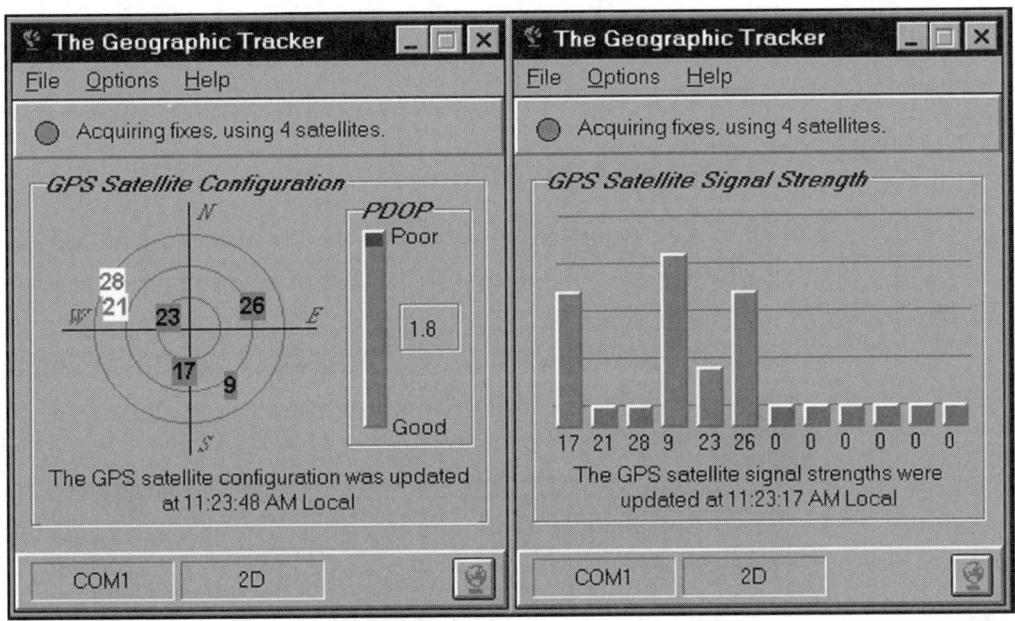

GPS satellite status display in Geographic Tracker.

Other Uses

The Missouri Department of Conservation developed a system to use GPS and mapping technology to manage an impending deer herd problem near developed areas and airports. With these tools the department can perform telemetry tracking and triangulation to map the ellipses formed by the movement habits of radio-collared deer and other animals. AT&T has traditionally used GPS and mapping to support quality control for designing long distance underwater cable routes. Blue Marble products are also used in precision farming, forestry, facilities management, vessel tracking and navigation, and telecommunications.

Routing Logistics

MapInfo's routing capabilities are not dependent on GPS technology. Businesses not yet equipped with GPS can still organize routing and delivery efficiently using MapInfo.

Tracking Utility Repairs

One Texas utility uses MapInfo to maintain data on high-voltage transmission lines and transfer stations. Field inspectors enter relevant data into their laptops and the information is saved in a database accessible to engineers who dispatch maintenance crews.

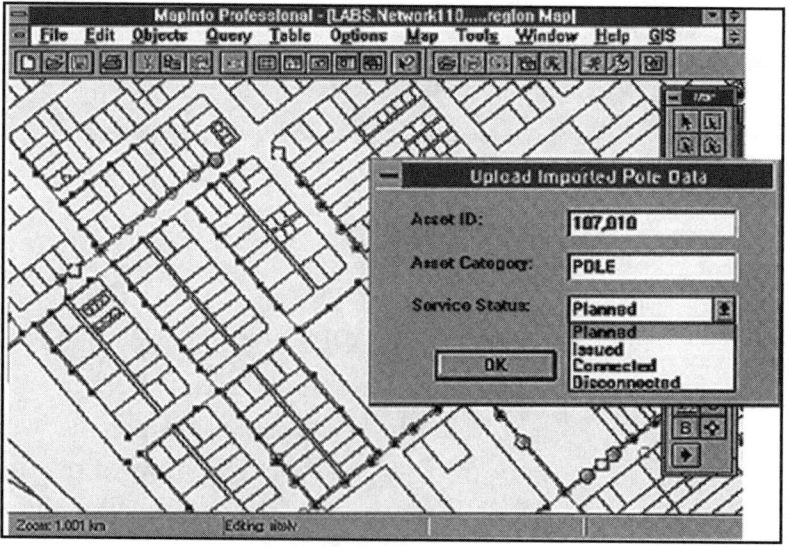

Utility use of MapInfo.

MapInfo cuts repair costs by decreasing the time between inspections and service. Before implementing MapInfo, up to two years elapsed between the inspection and maintenance, making most of the repair data obsolete.

Package Pickup and Delivery Solutions

MapInfo applications allow courier and package delivery businesses to organize planned pickups in relation to distribution facilities and optimize routing to match workload and package volume. Several third party packages contain routing tools to find the shortest path between two points, determine optimal delivery paths to multiple stops, and display roads within a given distance of the point of origin.

Finding the shortest path between two points is the simplest add-on function. As shown in the following illustration, this function simply computes the shortest distance from one spot to another.

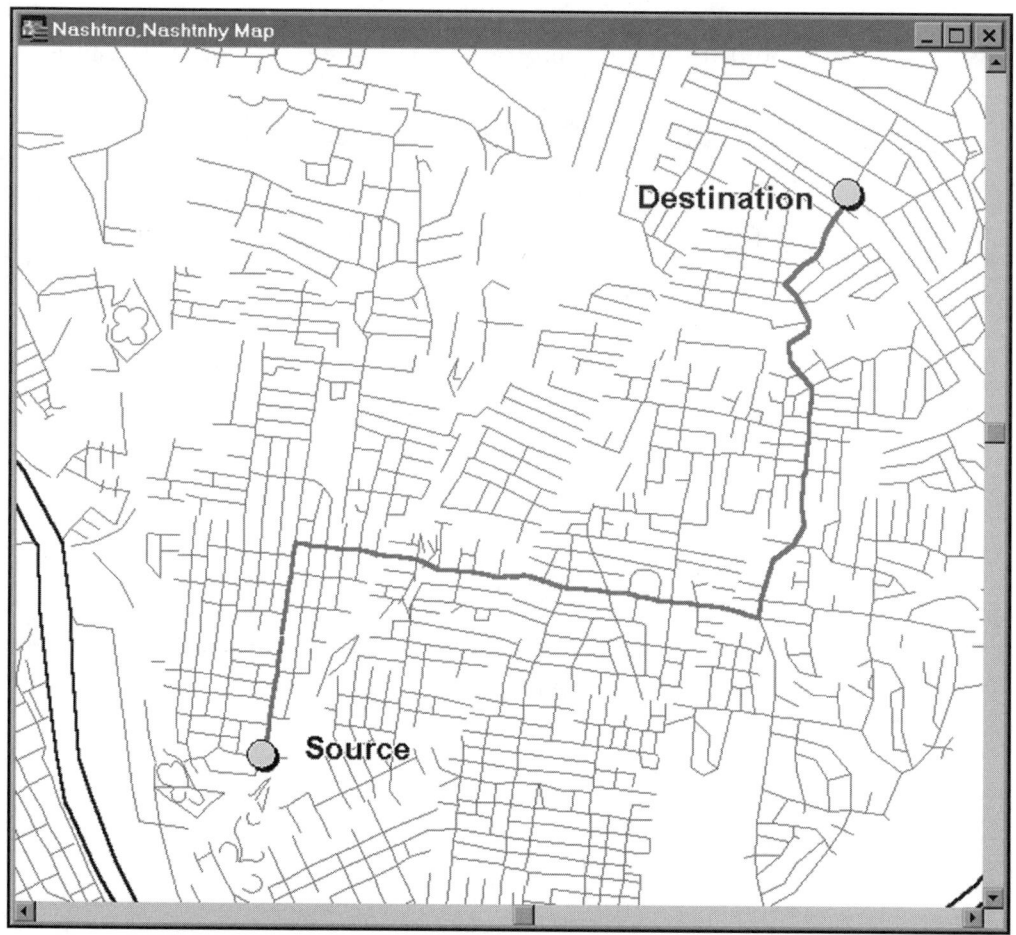

Shortest path between two points.

More sophisticated tools also take into account changes in terrain, road obstructions, speed limit, weather, and time of day—details that are essential for 911 dispatchers. (For more information, see the "Emergency Medical Service Institute" case study in Part Two.) Routing applications can also generate maps to show the streets and distances that will be traveled.

Routing Logistics

Next, routing tools can coordinate the movement of packages or products among several sites. To accomplish this, addresses are geocoded and a clustering algorithm is used to order the stops most efficiently. The result is a map of the best route for reaching all points. If necessary, directions and mileage can be added to the map and printed for the driver.

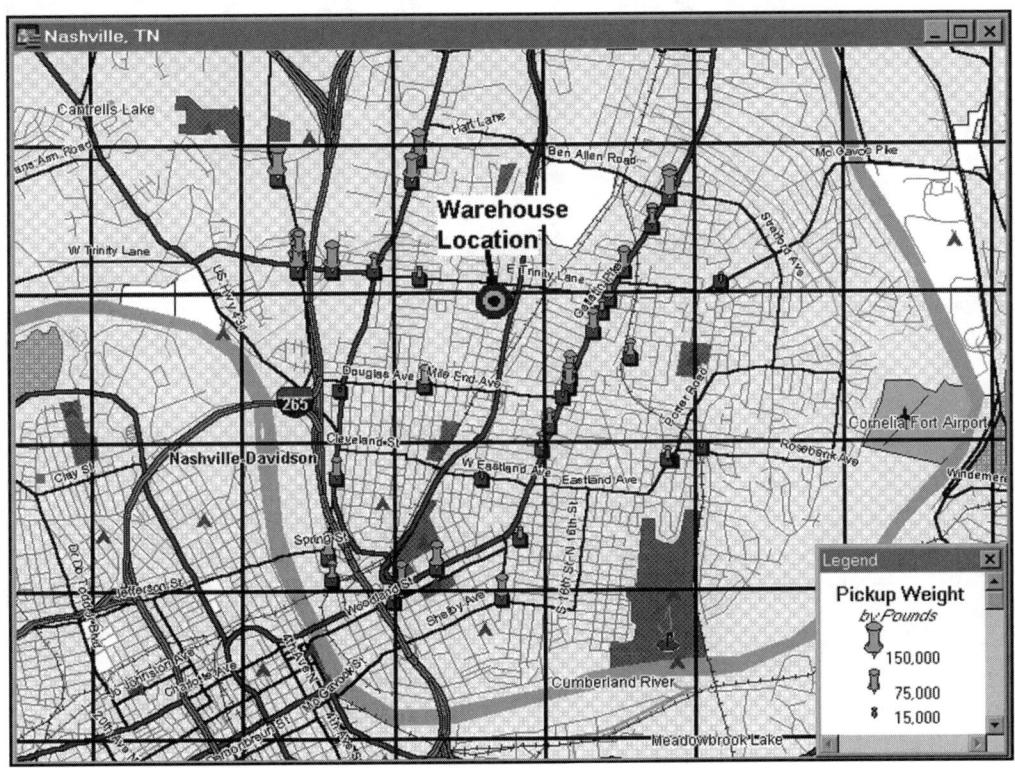

Size of pickup load.

Couriers, Inc
Daily Work Order
06/16/1997

Driver: Doug R.
Truck: 40' Tractor Trailor

PICKUPS	TOTAL WEIGHT	TOTAL VOL	TOTAL UNITS
Electra Sport	125	50	750
Flex Communications	650	175	1300
Jack's	1250	125	500
Friendly's	1000	75	100
Philey's	1250	110	250
Monroe Acoustics	1000	75	5000
Hersheys	850	400	10000
Hardware Express	1500	125	150

Pickup schedule report.

MapInfo can be applied in a similar fashion to school bus routing. Because multiple buses are involved, MapInfo's redistricting capabilities also come into play. The following illustrations show an initial routing scenario and the result after a routing algorithm and redistricting are applied.

Routing Logistics

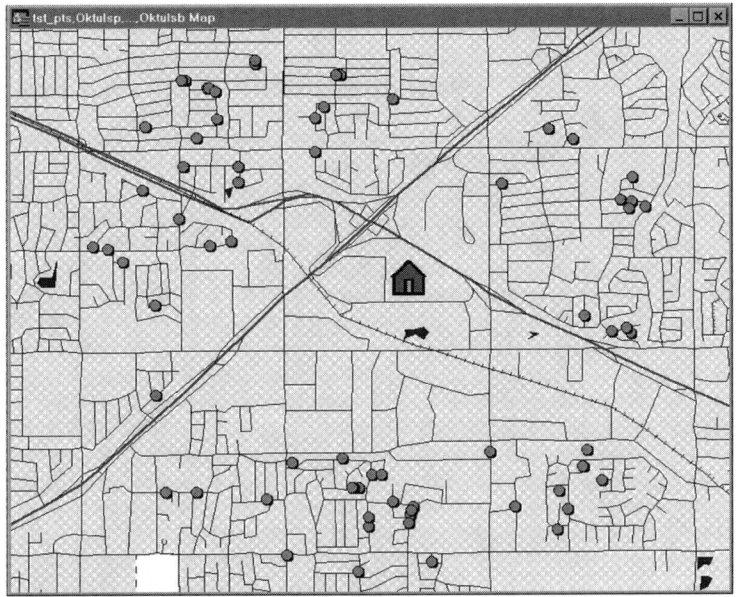

Plotting students as points on a map.

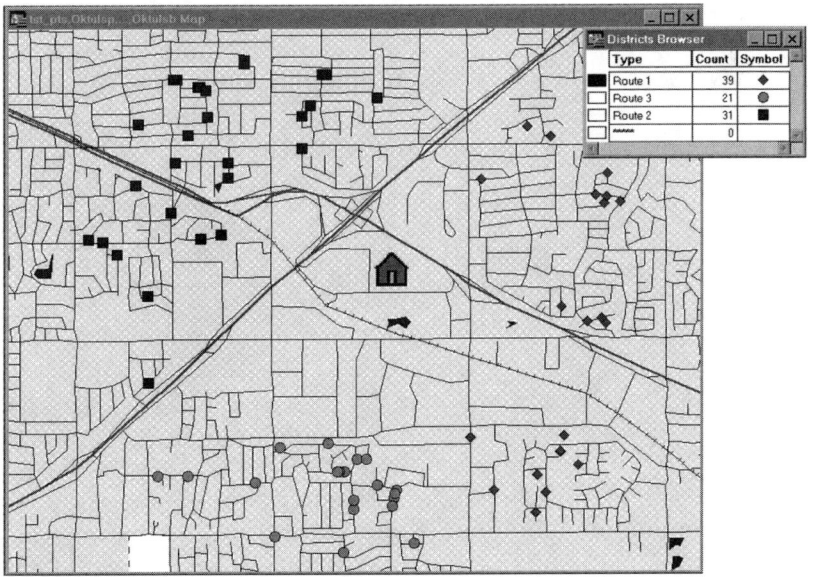

Initial districting of students according to the symbols shown in the Districts Browser window.

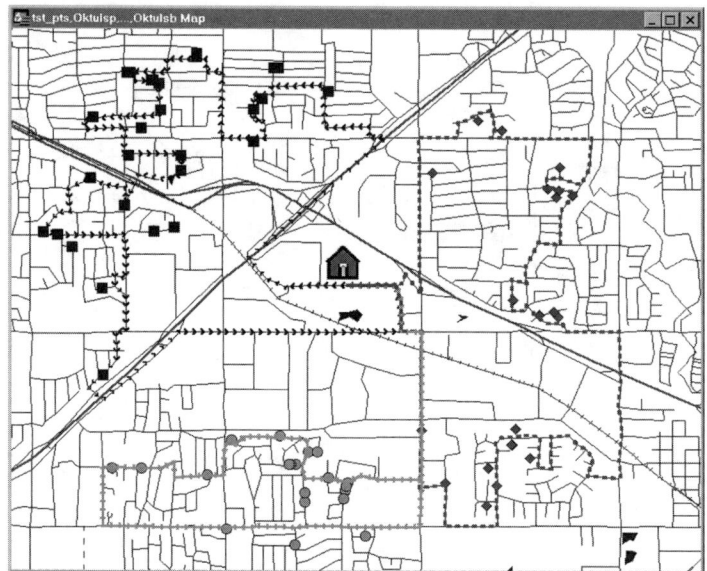

Balanced districting of students and bus routes.

3

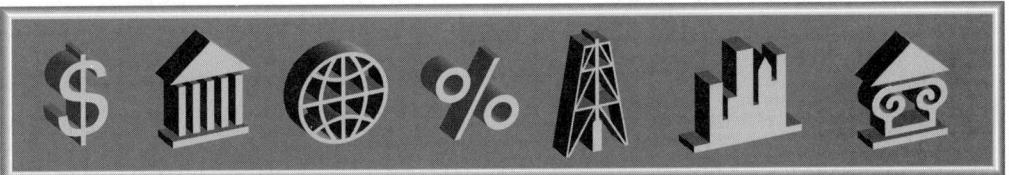

Telecommunications and Information Systems

Telecommunications

One of the fastest growing industries worldwide, telecommunications is changing profoundly as it becomes increasingly competitive. As regulatory control is relinquished, telecommunications players are positioning themselves to vie for millions of business and residential consumers. The competitive environment holds substantial promise for new market entrants and broadens the products and services historically offered by mainstream telecom companies.

While the deregulated market creates tremendous opportunities, it has also raised the competitive stakes enormously. Telecom companies are feeling pressure from stockholders, boards, customers, and competitors alike. In many ways, technology is driving the revolution. Companies who fail to keep up with advances cannot compete or serve their technology savvy customers. Worldwide, Asia and the Pacific Rim constitute the single largest telecom market. Within five years, the investment in telecom will exceed $300 billion worldwide. With such high stakes and profit potential, competition is fierce and bloody. Perhaps ironically, the competition is most apparent at the family dinner table—the time when residential customers are bombarded by telemarketers to change their long distance providers, try new services, and in some cases are promised cash to switch providers.

Within the past few years, telecommunications has evolved to encompass much more than local phone service. Mobile phones, cellular phones, and sophisticated PCS (personal communication service) phones with other advanced functions are flooding the market. In addition, telecom companies are offering Internet services and long distance packages.

The telecom industry is a natural fit for GIS because of the wealth of geographic data and the industry's need for up-to-date information. As such, it would be accurate to say business mapping providers are drawn to the industry like sharks to a wounded fish. Typically, companies operating in industries booming with emerging technology, new markets, and an infusion of cash require immediate and effective tools to maintain a competitive edge in mission critical areas. In telecom, GIS is applied to support marketing and sales, customer service, network management and planning, tracking buried cables, and competitive analysis. Each of these areas is explored in the sections that follow.

Marketing and Sales

Until deregulation, selling phone services required little effort and less research. Mobile phones were unheard of and most customers had no choice over their service providers. Today, bundled services and new products—from cellular and PCS phones to Internet providers—and a multitude of long distance choices have clouded the market significantly. Telecom companies are driven by two strategic goals: finding and attracting new clients, and maximizing and maintaining an existing customer base.

New Customers

As in many other industries, telecom sales and marketing professionals use GIS to pinpoint sales efforts. Companies rely on demographic and lifestyle segmentation data to target marketing approaches and match customers with appropriate products and services. Bundling enhanced services such as Internet access, three-way calling, caller ID, and other services for targeted customers can dramatically improve revenues and profits with minimal effort. ZIP Code and block group data can be used to locate customers in particular occupations or income groups, as well as identify concentrations of small businesses requiring specific telecommunications services. Identifying and marketing to customer bases with the best profiles improves the chances of reaching the right customers and reduces marketing costs.

Current Customers

It's not uncommon for many residential customers to switch phone companies over the course of a year. To prevent this from happening, telecom companies market to existing customers by "backing into" the customer bases most likely to yield the best profits. Telecom companies back into markets by analyzing existing customers to identify social and economic patterns typical of the most desirable customers for their products. Once companies quantify the qualities of their strongest customer base, they focus marketing efforts on finding similar customers. Monthly phone bills—particularly of large customers—contain ample usage data, and the geocoding process generates maps that uncover patterns and trends. Using this technique, telecoms begin to profile high end, high return customer bases according to social and economic characteristics. Once a profile is built, telecoms can market to similar customers throughout the market. Ultimately, having many customers is not as important as having customers who spend more money on a company's products and services.

Telecommunications

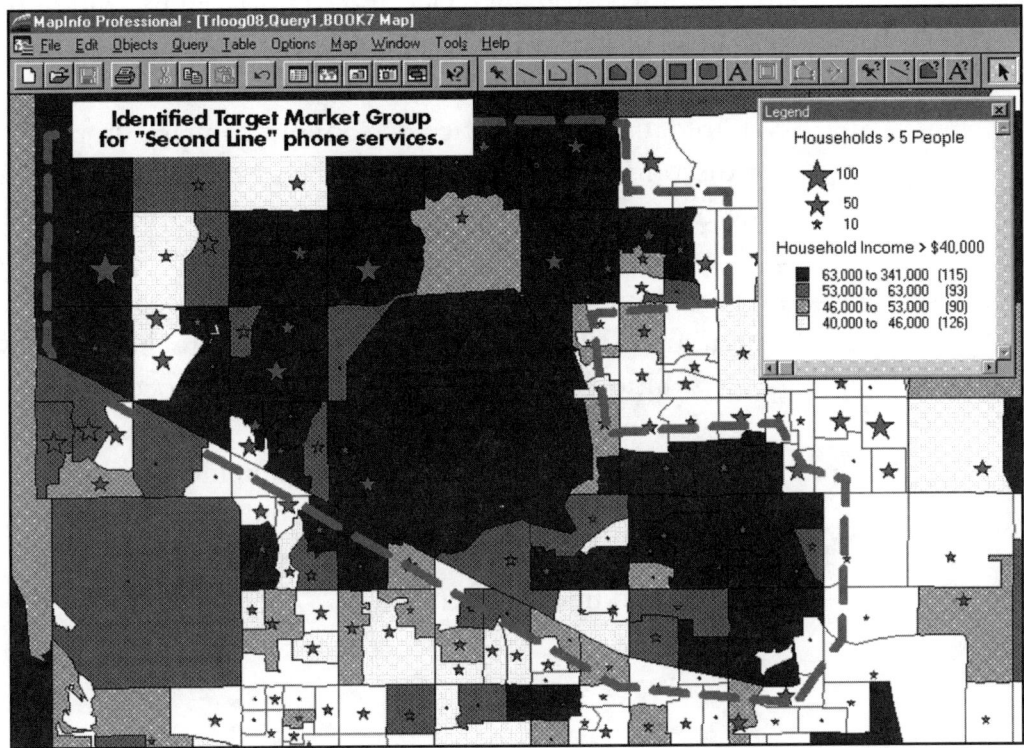

Backing into a market through geocoding and analysis of usage patterns.

Customer Service

No doubt the telecom wars have made customers more savvy regarding technology. In the initial race to attract customers, advanced products and services flooded the telecom market. Consequently, consumers have come to expect and even demand that superior products and customer service be standard. Satisfying a fickle customer base can be difficult—if not impossible—without adequate customer service tools.

Integrating customer service and operational information is critical when trying to reduce costs and improve network performance to satisfy real time customers. Using MapInfo, customer service personnel can map "dropped" (or disconnected) calls and other customer problems as they happen by geocoding the location of the problem and logging details in a MapInfo database. Once the problem is mapped, analysis can take place regarding the extent of the problem, which may lead to better and faster solutions. While disconnections will occur, mapping can provide immediate insight into areas with repeat problems that may indicate equipment failures or a shortage of necessary facilities. Such data can be shared with network engineers to make appropriate repairs, changes, and improvements.

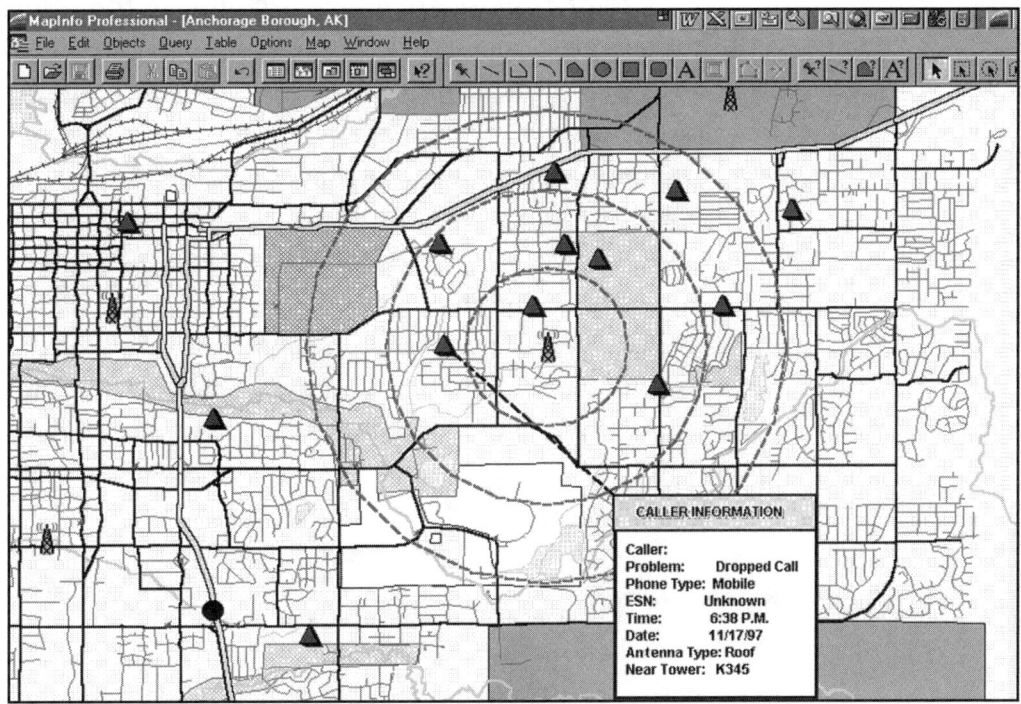

Mapping telecom operational problems.

MapInfo's capabilities with regard to customer service are virtually too numerous to mention. As the case studies in Part Two highlight, MapInfo may be considered by a company for one application only to be used to satisfy a number of subsequent needs as well. Indeed, desktop mapping applications are so diverse it can become difficult for users to remain focused.

Network Management and Planning

Customer service and problem tracking are only half of the network management equation. Projecting traffic growth, locating customer equipment, and the increasingly important issue of rural access comprise the other half. If a mapping system is being fully incorporated into company operations, system engineers can use mapping technology to layer an assortment of data for traffic studies. For example, engineers can layer routing and equipment location maps with problem occurrence maps to determine network deficiencies and make decisions regarding improvements in capacity or revisions to existing routes. Projecting growth is critical to ensure that the network will continue to adequately service the company's customer base. MapInfo can map areas with growth trends so that the correct routes and capacity can be properly incorporated into the network.

Rural Access

One critical element of telecom reform is rural access. As part of the deregulation process political leaders mandated that all residential customers—including those in remote areas—have access to the technological advances driving the industry. However, making that mandate a reality for rural customers is expensive and difficult. The competitive nature of the business impacts how rural areas are incorporated into each company's strategic plans. MapInfo can help companies develop effective and efficient *visual* plans to carry out the mandate, including analyzing current facilities, future capacity needs, and how future rural lines can be incorporated into the company's network. At present, several MapInfo partners are supporting the development of rural networks by attempting to mathematically model rural areas for potential capacity and demand.

Tracking Buried Cables

It would surprise many people to learn the amount of utility technology stored underground. So many cables are stored below ground that digging a hole without sufficient information can be expensive and dangerous. Millions of dollars are spent each year repairing utilities damaged by accidental digging. Desktop mapping applications that address buried cables—also known as "call before you dig"—help maintain customer service and safety. They also can help coordinate upgrades and expansions in network carrier technology such as the installation of fiber optic cable. MapInfo can layer network line maps, streets, railroads, water systems, and electrical, gas, and cable lines so that potential digging locations can be assessed and future problems averted. As more lines are laid underground, providers of diverse utilities may even share geographic line data to eliminate problems and prevent accidents.

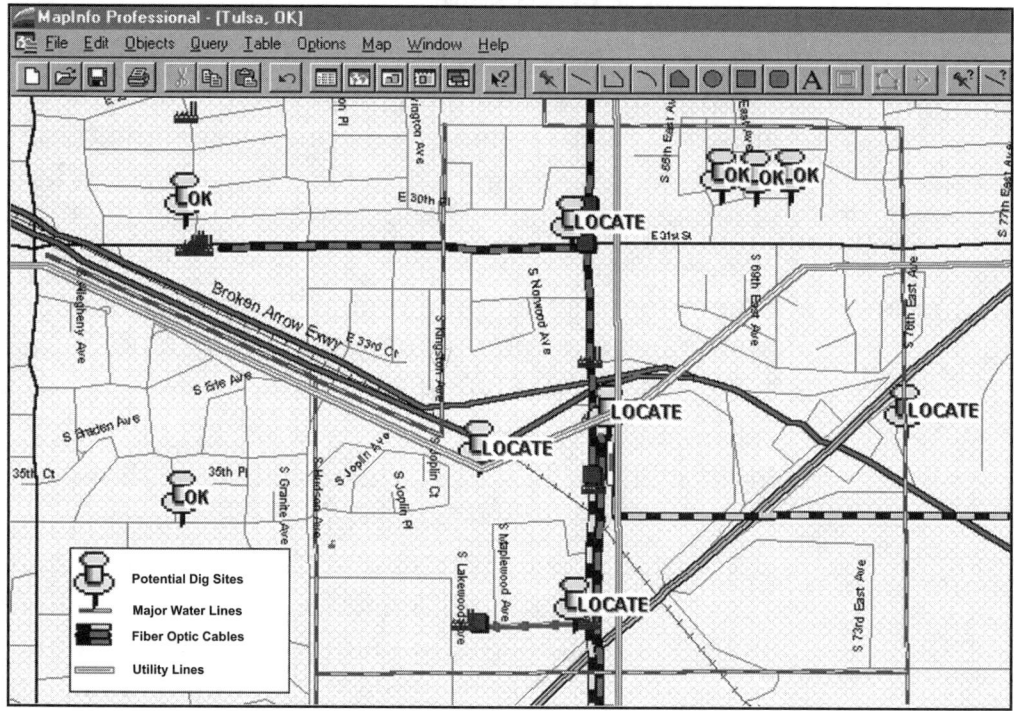

Sample MapInfo map showing underground utilities.

Competitive Analysis

In every competitive endeavor, it is essential that a company understand its own strengths and weaknesses, as well as those of its competitors. Because the majority of telecommunications data is visual or geographic in nature, competitive analysis frequently employs mapping. MapInfo can be used to track competitors' service capabilities, network availability, market coverage, and marketing tactics to shape a company's strategic sales initiatives.

Deregulation has also made it easier for new competitors to enter the marketplace. The explosive growth and competitive nature of the industry has fostered and will continue to generate mergers and acquisitions. Mapping applications can be used to assess synergies and redundancies between merger participants and evaluate the cost of the business venture.

Telecommunications Data

More than any other industry, telecommunications enjoys an outstanding volume of third party mapping data. Available data can range from more generic street and boundary files to specialized databases developed specifically for the industry. See Appendix B, "Reference Material and Data Sources," for more information.

Information Systems

While MapInfo applications are diverse, most customers also use the software as a basic informational database in addition to its other applications. An effective warehouse for storing and accessing a wide range of company data, MapInfo's strengths in this area are so profound that when a company sees the data presented in MapInfo, the software generally sells itself. For most potential MapInfo users, visualizing business data on a map for the first time can be a revelation. Once users understand the basic capacity of mapping technology—visually displaying data on a map—they quickly venture into more sophisticated uses and applications. Some of the more unusual and distinctive applications are discussed in the sections that follow.

Intranet/Internet Applications

Many businesses have learned the value of hosting Web sites. At present, one way to distinguish a Web site from a sea of competing home pages is to "spatially enable" the site. In essence, this means applying mapping technology to the Web site. Typically used by retailers, the approach enables a Web customer to type in his or her address and obtain the location of the retailer's closest store. The site may also display a map showing distance to the store and offering key data such as address, phone number, and hours of operation. When Internet mapping technology was introduced in 1996, it was prohibitively expensive. Since then, the cost has dropped dramatically to allow even modest Web sites the opportunity to benefit. The MapInfo Reseller Locator site appears in the next illustration.

Information Systems

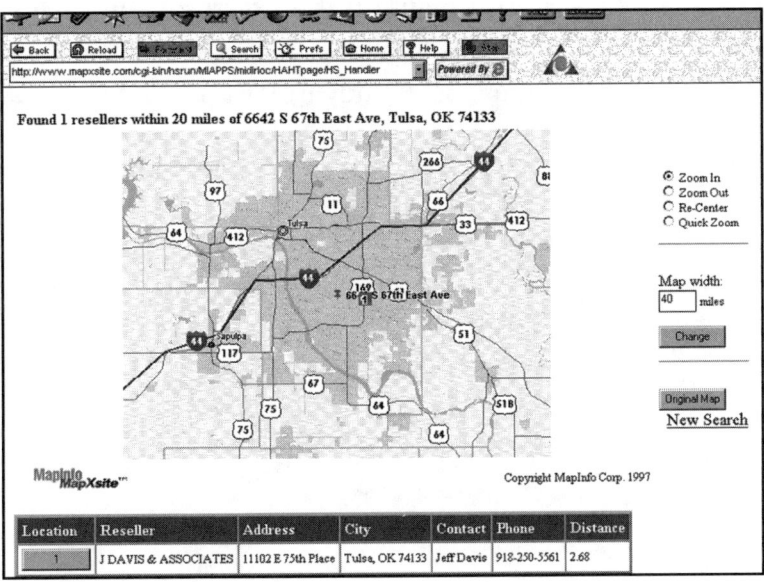

MapInfo's spatially enabled Web page.

Oil and Gas Well Mapping

Mapping applications are increasingly developed for the oil and gas industry. One particularly useful application, created by IntelleVue, manages oil and gas well production information in order to generate 9-section plat and production maps. In an industry of sophisticated, expensive geographic software, MapInfo bridged the gap between this expensive software and multitudes of different end users. IntelleVue developed a system that merged industry standard well databases and simple mapping tools to access key well data and print valuable well production maps. In addition, because MapInfo is so cost effective, many more users can access the software and print maps on their own schedules. For more information on this application, please see the "Louis Dreyfus Natural Gas" case study in Part Two.

Information Systems

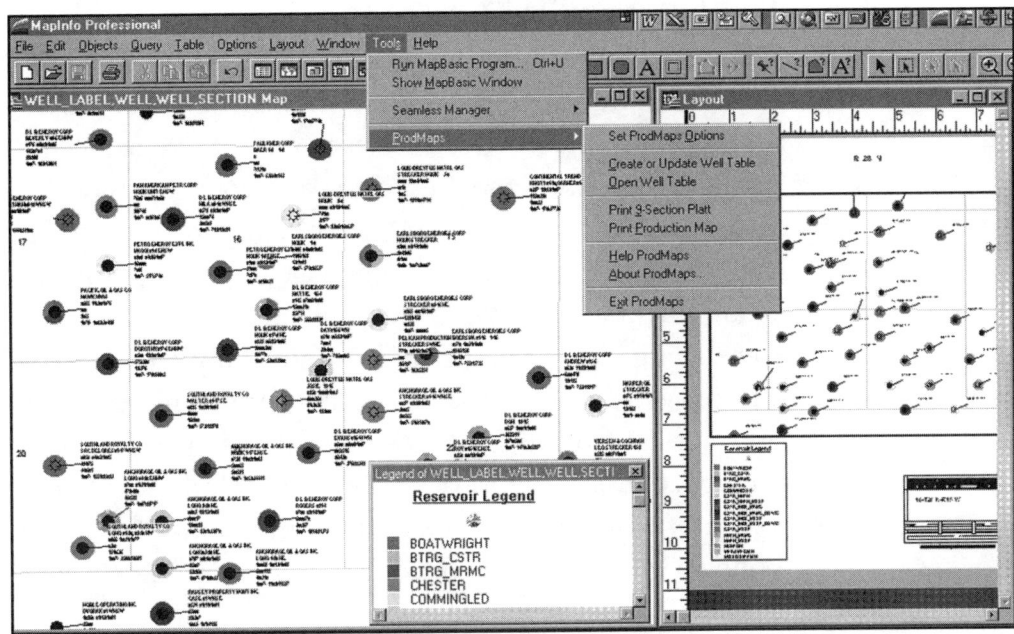

An oil and gas well mapping application in MapInfo.

Accounting Management/Reporting

Another unique mapping application is displaying monthly accounting reports. Managers and executives can access online mapping data to focus on accounting details instead of poring over spreadsheets and printed reports. Trends in profitability, sales, and costs are easily displayed at virtually every level of detail. Viewing profit and loss statements and sales revenues using a mapping system makes the data more accessible, fosters understanding, and provides a more dramatic method for illustrating key factors and trends. The information, updated on a timetable, is available the same time each month.

Human Resources

Large geographically diverse companies can also benefit from mapping human resource information. Identifying and grouping health care services and programs according to geography is more efficient and cost effective than traditional approaches. In addition, mapping offers a rare opportunity to proactively manage equal opportunity requirements by geography. Business geographics software using demographic and ethnic mobility databases can provide benchmark equal opportunity requirements and act as a tool for reporting such data. Geocoding employee home addresses using MapMarker can help executives analyze the political ramifications of company policy within a specific region.

Politics

Managing political information, including geocoded registered voters, geocoded donors, and previous election results, can be invaluable for elected officials and candidates. However, the fundamental building block for a political GIS—digitized voter precincts—typically does not exist. Most municipal election boards store such data as magic marker outlines (or even crayons) on crude paper maps. To overcome this problem, IntelleVue partnered with a political consultant and digitized every voter precinct in Oklahoma from county election board maps. The project yielded powerful results for elected officials and candidates.

Desktop Mapping in Brief

Hopefully this chapter has helped to increase your understanding of the breadth of desktop mapping applications in telecommunications and information systems. The state of GIS today may be comparable to the early years of spreadsheet use. Initially, spreadsheet users were interested in the different ways information could be displayed. However, as the technology advanced, users began to use spreadsheets as data management tools that revolutionized how business information was stored. While having a similar impact, business geographics may offer an even more powerful technology because it combines robust analysis with high quality visual presentation.

4

Banking, Insurance, and Real Estate

Banking

These are confusing times for bank managers and analysts. Mergers and acquisitions are forcing banks to change ownership faster than professional ball players switch teams, and bank customer bases are evolving as well. In addition, the explosion of automated teller machines (ATMs) and banking via the Internet have altered the profile of many banking customers.

To accommodate these changes, banks are streamlining operations, rationalizing retail networks of bank branches, and implementing procedures to manage a growing and diversified customer base more effectively. In particular, the change in customer base brings enormous amounts of information into play. Data on loans, deposits, interest rates, account numbers, customer lists, transactions, and monthly statements must be processed in order for banks to remain competitive. To fully develop customer relationships, bankers must take control of these data and maximize their usefulness. Access to customer data also enables banks to abide by federal fair lending laws such as the Community Reinvestment Act (CRA) and Home Mortgage Disclosure Act.

Mapping to the Rescue

The days of underpaid and overworked bank analysts who must pore over stacks of paper reports and spreadsheets armed with calculators are fading. Desktop mapping tools are revolutionizing the analysis of banking information by converting the process to an accessible geographic format. Consequently, maps with meaningful symbols and differentiated colors have turned reams of raw data into information that can be quickly analyzed and digested.

As analysts absorb the value of GIS, the GIS rallying cry is growing louder. Banks increasingly understand that data can be layered to create "pictures" that uncover new patterns, trends, and opportunities. PC based mapping solutions are used for all aspects of banking, including network optimization/site selection; CRA compliance; and marketing, advertising, and customer analysis.

Network Optimization/Site Selection

With the inception of interstate banking, the industry has become a never ending story of mergers and acquisitions. Banks that survive this market realignment must make smart, quick decisions regarding their retail networks. Acquisitions and mergers create overlapping branches and duplicate services. The odds for long-term success are enhanced for organizations who make decisions to improve customer service. Properly servicing clients involves refining the bank's mix of offerings; rationalizing branch facilities to minimize overhead and maximize market share; correctly siting new ATMs; and effectively implementing home banking technology in appropriate areas.

To accomplish the above tasks, banks must thoroughly understand respective customer bases in order to determine the best approach. Merging banks can geocode and then match both sets of customer databases according to residential and business addresses. This process identifies duplicate customers and highlights new customer concentrations that can be assessed as part of channel analysis or target marketing. Facility overlap is another by-product of mergers and acquisitions. Mapping facilities against customer concentrations can pinpoint areas with insufficient, adequate, and duplicate facilities. It also allows users to analyze specific locations more closely.

Visualizing key market elements (e.g., existing customer base, facility network, market demographics, market street network, traffic counts/flow, and competitor networks) enables planners to quickly evaluate competitive environments, rationalize branch operations, and develop new sites.

CRA Compliance

Historically, the banking industry has relied on maps and approached markets geographically—although in a very different manner than used in business mapping. Bankers used wall maps to identify "risky" neighborhoods for loans and denied loans to certain customers based on their addresses, not credit worthiness. While it is good business practice for banks to minimize the number of high risk loans they issue, some loan decisions were being made according to geography, not an individual's application. To prevent loan discrimination, the federal government passed the CRA in 1977, which required banks to establish mandatory market areas for service coverage—70% to 80% of a bank's overall loan portfolio—and provide a reporting structure to verify compliance.

In 1977, GIS was still an expensive option, and bankers continued to stick pushpins into wall maps. However, as PCs and business geographic software became more prevalent, bankers were able to use computers to track loans and define and optimize service areas. Historically, bankers have determined service areas and loan percentages on a branch by branch basis. Bankers would draw circles of specific sizes around branch locations and then compute loan percentages, rather than the reverse (i.e., using geographic loan distribution to determine appropriate service areas). This process was completed on all branches until a total market area was defined. Unfortunately, this process often produced service areas that encompassed or exceeded entire metropolitan areas, yielding an unrealistic picture from both business and customer service perspectives. Unwieldy service areas left banks open to criticism that they were not adequately serving or providing loans to certain neighborhoods.

Desktop mapping tools offered a solution to this problem. One of the most up-front and defensible solutions employs the three-step process below.

1. Geocode loan accounts to map locations of loan recipients.

2. Assign loans to the nearest branch (regardless of where they originated) to determine each branch's market area.

3. Create the total bank market area using the data assigned under Step 2.

The above process allows bankers to identify and label CRA market areas more accurately. As the following illustration indicates, CRA market areas determined using desktop mapping techniques are dramatically smaller than areas defined via the above mentioned process. Better defined market areas also allows bankers to focus their efforts on market demographic profiles and ultimately improve service with additional products and loan packages. A third major benefit of business mapping tools is their versatile presentation functionality, which can bolster a bank's credibility with regard to CRA compliance reporting.

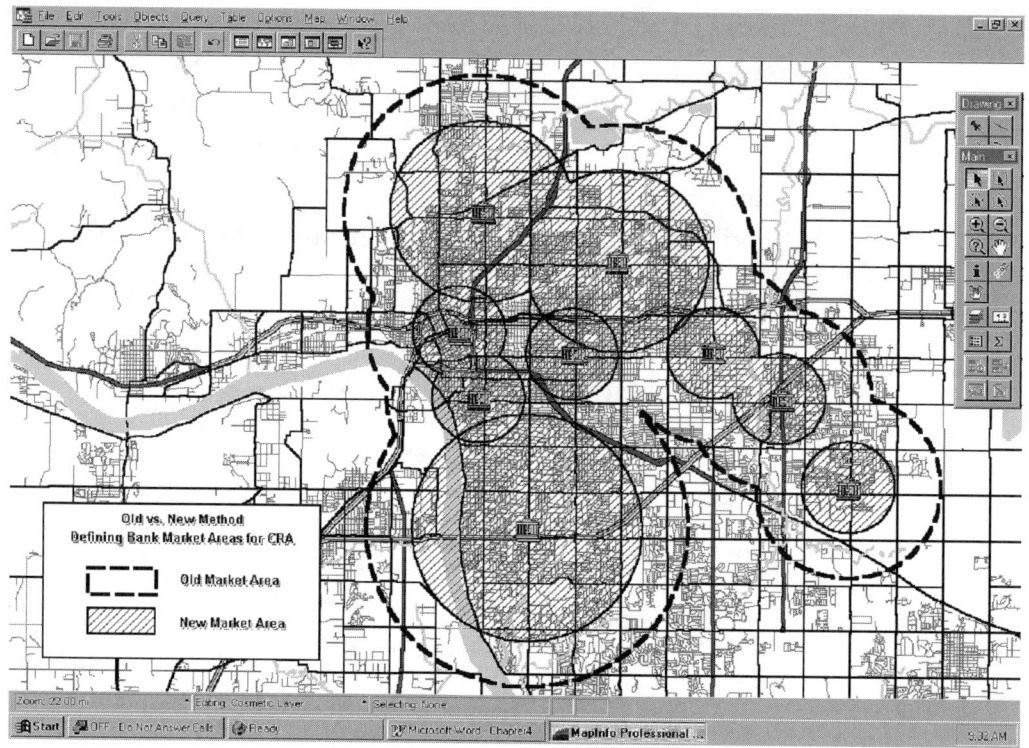

Old (outer circles) versus new (inner circles) CRA market areas defined using MapInfo.

While MapInfo Professional can support the CRA function off the shelf, several custom CRA applications have also been produced by MapInfo partners (also known as value added resellers, or VARs). A list of such applications is available by accessing MapInfo Corporation's Web site at *www.mapinfo.com*.

Marketing, Advertising, and Customer Analysis

MapInfo's ability to target customers according to specific products and services can be a tremendous tool for advertising and direct mail campaigns. These can be extremely focused when customer, demographic, and ZIP Code boundary files are layered together. Business mapping reduces advertising costs and improves responses because only relevant customers are targeted.

According to a MapInfo banking white paper (a marketing technical document), a community bank in a metropolitan area targeted individuals older than 50 years of age and earning more than $50,000 annually as likely candidates to open certificate of deposit (CD) accounts. MapInfo enabled marketers to query the market demographics for census block groups with a high concentration of prospective customers. Next, they compared the existing geocoded client database and eliminated individuals who already had such accounts. Mailing lists for the targeted ZIP Code areas were purchased. To supplement the direct mail campaign, the bank advertised in selected local newspapers instead of major daily papers, which significantly reduced advertising expenses. Finally, the bank tracked results through MapInfo. To measure success, the bank geocoded the new accounts and compared the results to the initial target area. The resulting map confirmed the efficiency of the target marketing campaign, showing that an initial investment of $6,000 yielded $3 million in new CD accounts.

Staying Ahead of the Game

Mapping will continue to be a tool of choice for the banking industry as mergers and acquisitions persist, marketplace demographics shift, and new technologies alter banking. The ability to visualize information and provide immediate access to data critical for intelligent decision making is a competitive advantage and mandatory element of banking in the 21st century.

Insurance

The insurance industry has been jolted in recent years by catastrophes of biblical proportions. Disasters such as Hurricanes Andrew, Hugo, and Iniki; the massive Mississippi Flood of 1993; and devastating California brush fires have caused billions of dollars in property and infrastructure damage. The liability and exposure for insurance companies are forcing many companies to become proactive disaster managers. Being responsive to customer needs and applying a proactive management style will likely influence the way insurance companies conduct business in the future, and business mapping can smooth the transition. While the most common use for mapping today is in underwriting, other growing applications include risk concentration analysis (RCA), disaster and catastrophe response, customer service, and sales and planning.

Underwriting

Mapping has become an irreplaceable tool to evaluate risk for insurers. In the past, risk evaluation relied on local knowledge and was fairly labor intensive. However, the volume of information used in such computations necessitated automating and streamlining the process. For example, calculations must take into account a constantly changing customer base (i.e., shifting risk patterns and liabilities) and key intangibles such as toxic waste sites, flood plains, and areas where criminal activity is prevalent—to name but a few.

Environmental Risk Assessment

Insurance companies can accurately determine the potential for risk if they can identify policyholder locations relative to potential disaster areas. The location of customers in high-risk areas such as flood plains, earthquake fault lines, tornado alleys, and hurricane paths can identify problems before they occur and allow insurers to properly assess the likelihood of future claims, thus assessing a fair price.

The list below describes historical data that are currently available to allow insurers to analyze the potential for future events.

- Weather/natural disasters (e.g., hail, wind, hurricanes, tornadoes, floods, and earthquakes)
- Environmental disasters (e.g., hazardous and/or toxic waste spills and sites)
- Artificial environmental risk areas (e.g., oil and gas pipelines, refineries, and nuclear power plants)

Automobile Drive Distances

Driving distance to work is a common variable for determining car insurance premiums. Industry surveys report that people tend to underestimate the distance they drive to and from work, which affects the risk calculation and can cause insurers to lose money. To counteract this possibility, some insurers collect home and work addresses and geocode them to a map to accurately determine probable minimum driving distance. The insured's proximity to high-risk intersections can also be factored into the calculation to assess the potential for future auto claims.

Choosing the Correct GIS Solution, Consultant, or Software Package

Businesses implementing a GIS face many potentially confusing options. Choices can include evaluating and selecting a software package off the shelf (e.g., MapInfo Professional, ArcView, or other products), purchasing a custom application, internally developing a product, and hiring a GIS consultant to assist with or complete the entire project.

The correct choice, which inevitably lies with the end user, depends largely upon the following factors.

- What end users must accomplish with the GIS
- How central the GIS application will be to company performance (also known as its mission critical nature)
- Project budget

Answers to these questions will offer insight into the correct solution. However, end users must take care to avoid embarking on a path that could turn the project into a disappointing experience. Regardless

Insurance

of your situation, obtaining expert advice is likely worthwhile. Within the MapInfo marketing channel, a network of authorized or strategic partners can assist you with this decision. Partners will strive to understand your situation and propose the most effective and feasible solution, whether that means purchasing a single MapInfo Professional license or developing a customized system. Because the quality of your mapping system depends upon data quality, every comprehensive solution should involve obtaining data necessary for your applications.

If you decide to involve a GIS consultant, the following guidelines can help you evaluate candidates.

- Check references to obtain input from other customers regarding the consultant's work. Ask for a list of all clients over a specific time period, not merely a list of references.
- Ensure that the consultant's work is guaranteed.
- Demand that the structure of the project include joint development review sessions.
- Make sure that you are informed of all up-front costs, as well as costs that will be incurred throughout the project's duration.

A competent GIS consultant will understand your concerns, anticipate your needs, and provide the most appropriate solution—not just sell you a software package and then move on to other clients. Many GIS projects evolve, and your relationship with a consultant may last for several years. Make sure that you create a strong foundation at the outset of the project.

Risk Concentration Analysis

The magnitude of recent disasters has compelled insurers to gain a more complete understanding of the distribution of their customer base in order to alter portfolios and avoid major losses resulting from a single disaster. Insurers have incurred substantial losses from onetime disasters when many customers were clustered together and their property destroyed. Known as risk concentration analysis, this process is the second most common use of mapping technology for the insurance industry.

Risk concentration analysis quickly shows where customers are concentrated. If a large percentage of an insurer's portfolio is within 0.5 mi of the San Andreas Fault, the next major earthquake could take the customers and the company with it. Insurers must minimize their exposure to potentially devastating disasters—particularly if they are not as rare as anticipated. Along with clustering studies, mapping is also a useful tool to perform probable maximum loss (PML) analysis, which establishes a relationship between an area's total PML and its proximity to potentially dangerous environments. Ultimately, the proximity of dollars insured to potential disaster areas is perhaps more important than the number of policyholders who may be at risk.

Insurance

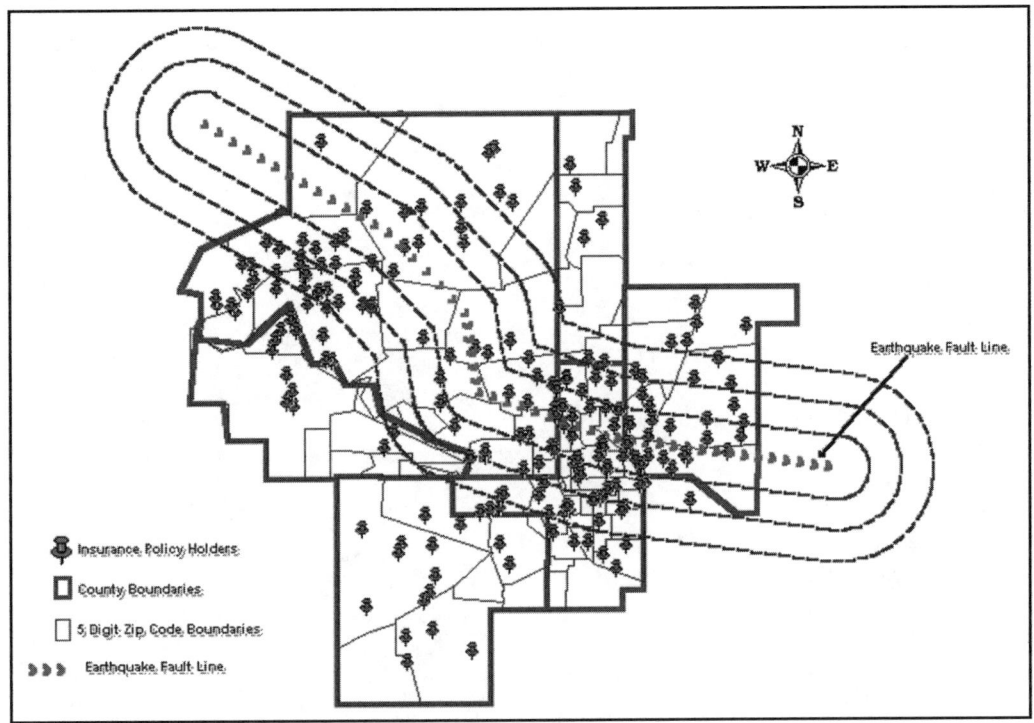

Potential liability analysis using an earthquake fault line.

On Target Risk Series

On Target Mapping provides a series of nine environmental, weather, and national disaster assessment databases. Compiled from historical events and digitized into a mapping format, these databases can be incorporated into insurance analyses to assess the probability of similar events in the future.

Sample maps generated by layering the HazWasteInfo database with two others offered by On Target. At top, hazardous waste sites near Bridgeport, Connecticut, shaded by volume of waste generated per calendar month. The middle image depicts the correlation between hazardous waste sites and, using the ToxicReleaseInfo database, toxic chemical release sites in Connecticut. The bottom image illustrates the relationship between hazardous waste sites and, using the HurricaneInfo database, estimated hurricane paths in the state.

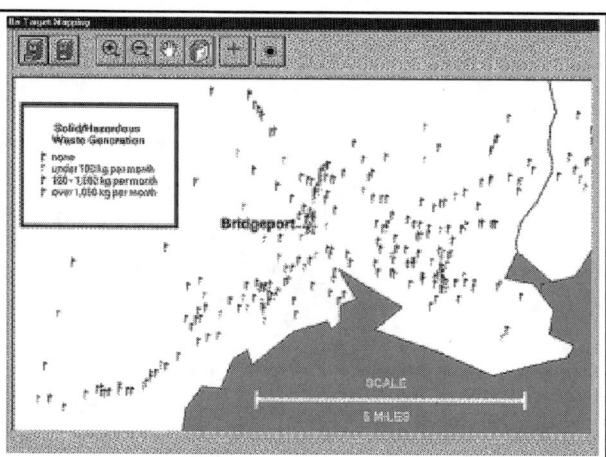
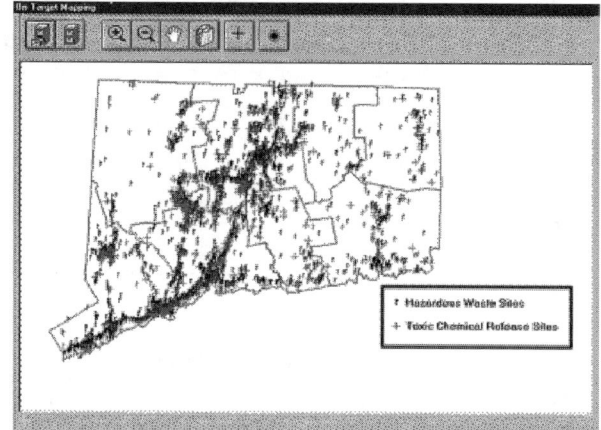
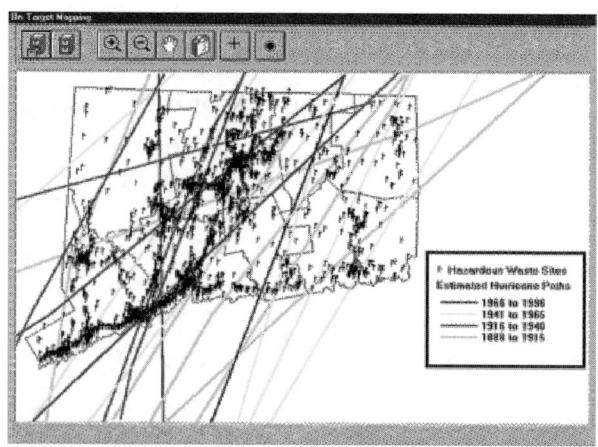

Disaster Planning/Catastrophe Response

The ability to quickly and accurately predict the cost of a catastrophe arrived only with the advent of GIS technology. Insurers can now realistically determine affected policyholders and the extent of potential damage and cost, sometimes days before certain types of events (generally hurricanes or floods) occur. This capability enables insurance companies to liquidate assets and generate only the necessary funds, saving millions of dollars. Other assets of mapping technology include predisaster planning and assistance, as well as customer service. If an insurance agent appeared at your door with a moving truck two days before a flood crested, your opinion of the individual (and the company) might noticeably improve. Assisting customers before known disasters hit can dramatically enhance the relationship and provide an opportunity to reduce potential loss.

In 1993, insurers used many aspects of business geographics to manage the Mississippi Flood. Real-time satellite images were imported as a data layer to track the extent and flow of water as the event unfolded over several days. In the next illustration, the dots represent each policyholder according to policy amount; darker values represent more expensive policies (see the Legend). The thick lines represent normal river level, the light gray area shows the flood plain, and the medium gray tone represents the extent of the flood at full crest, which was created using MapInfo's buffer tool. Together the layers provide a tremendous amount of information that can be absorbed and understood immediately, and the impact of the image far exceeds what might be conveyed through non-mapping applications. Using MapInfo, the insurer was able to identify each policyholder that was or would be flooded, the policy amount, and when the property would be affected. Consequently, many policyholders were able to move to higher ground before sustaining damage.

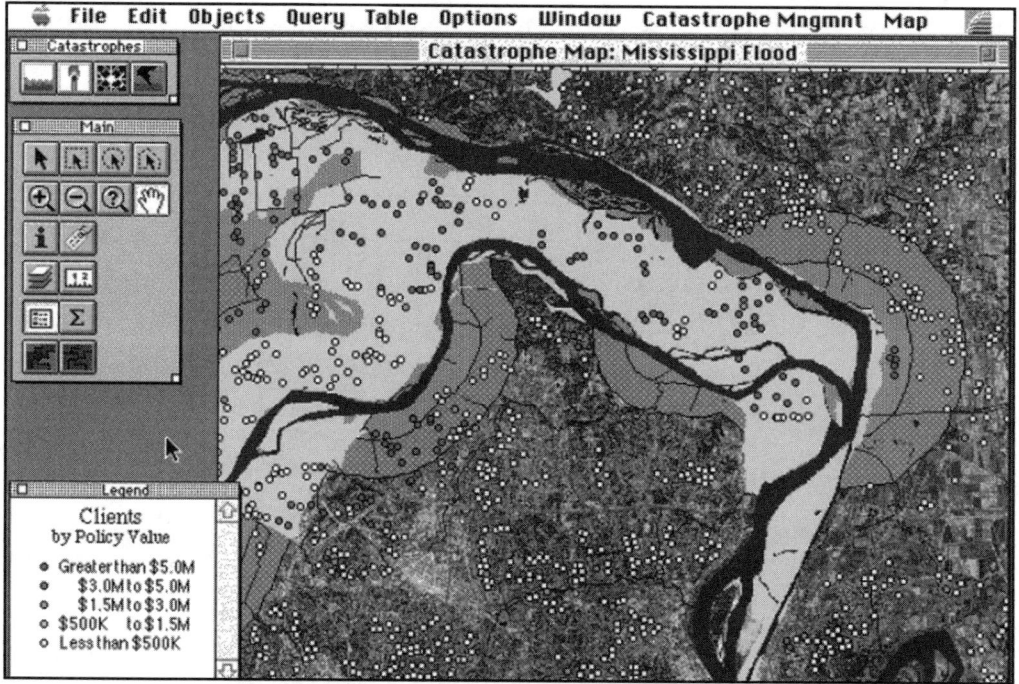

Policyholders in St. Louis, Missouri, affected by the Mississippi Flood of 1993.

Sales and Planning

The high cost of health care insurance is a common complaint among companies and employees forced to pay what they see as excessive premiums. In the competition to win large insurance accounts—the cost of which can reach $100,000 to capture a single account—mapping technology can be a distinct competitive advantage. The ability to portray to companies the locations of health care providers in relation to employees can often decide the issue. The next two illustrations reinforce the usefulness of mapping in such situations. While accurate, the table has little impact when compared with the following map. Moreover, the map confirms that the locations of health care providers offered by a competing company are less convenient. Converting routine statistical data into a colorful picture that is easy to understand may be the competitive edge required to close the sale. This approach can also be used to solicit new health care providers for the network.

Dr. Joseph E. Smith	W Broadway	Carnegie	Ok	93015	555-654-1008
Dr. Bill D. Thomson	720 S 4th St	Chickasha	Ok	93016	555-224-5361
Dr. Bill D. Thomes	7770 W Hwy 87	Duncan	Ok	93533	555-255-2771
Dr. S. Jelly Brown	2007 S Division	Guthrie	Ok	93044	555-282-6440
Dr. Henry V. Breshing	W Broadway	Guthrie	Ok	93044	555-282-6440
Dr. William W. Davis	720 S 4th St	Oklahoma City	Ok	93116	555-842-4774
Dr. Tom J. Broojs	7770 W Hwy 87	Oklahoma City	Ok	93112	555-642-0685
Dr. Hubert R. Burches	7778 W W 67th	Oklahoma City	Ok	93116	555-842-4774
Dr. Jent Celle	6077 W W 67th	Oklahoma City	Ok	93132	555-720-1599
Dr. Wilson E. Clerkin	7078 W E 76th St	Oklahoma City	Ok	93111	555-427-5555
Dr. John J. duPlessis	7400 W W Expressway	Oklahoma City	Ok	93112	555-942-0685
Dr. Paul E. McMartney	6768 W May	Oklahoma City	Ok	93112	555-843-6062
Dr. Phem Binh Gone	4607 W Classic	Oklahoma City	Ok	93118	555-840-9966

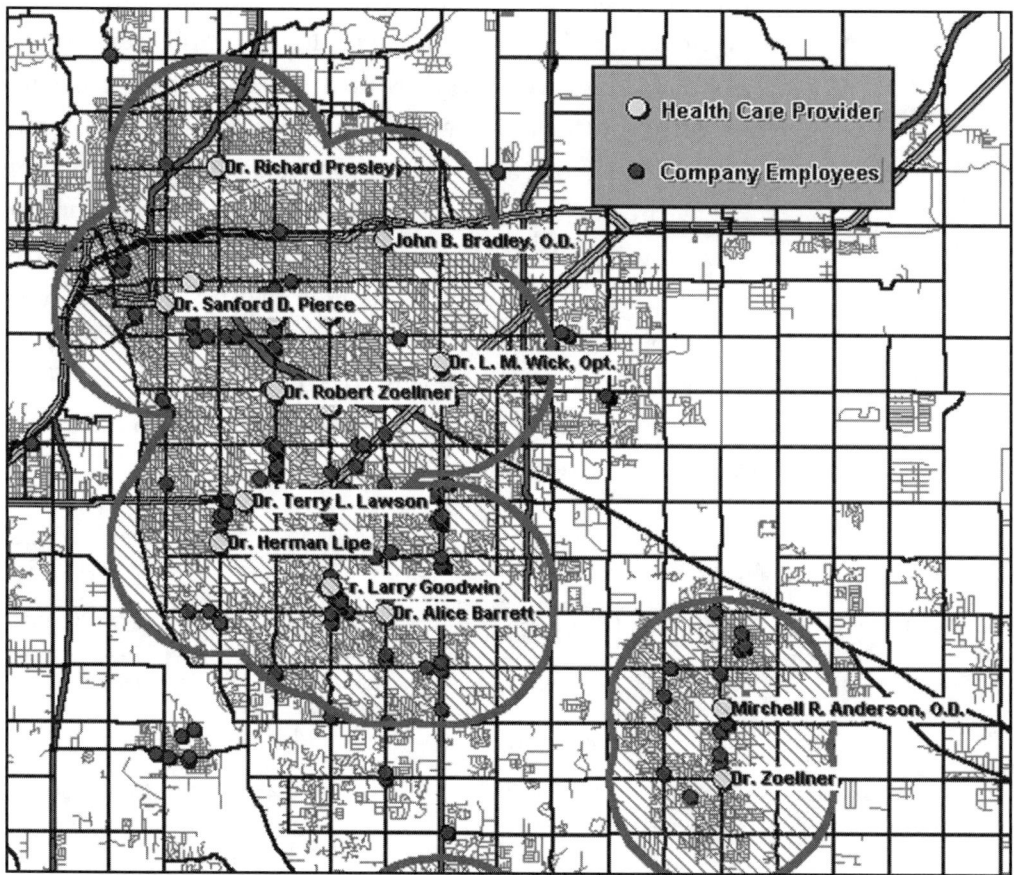

This map quickly demonstrates that 90% of the company's employees live within 2 mi of a health care provider on the network list.

Mapping has also dramatically improved the hit rates on direct mailings and other forms of target marketing. Reducing costs and improving sales can have a profound impact on profits and sales strategy. Profiling potential customers according to insurance needs and then targeting selected groups gives salespeople substantial leverage. Identifying potential customers according to income level, occupation, age, and other variables enables marketers to query a mapping demographic database and highlight census block groups matching the criteria. Salespeople can pinpoint customers who typically purchase the same product and match their insurance needs with the right products. This strategy lowers advertising and mailing costs and increases success rates.

Real Estate

Real estate represents yet another "natural" application of business geographics. Common real estate uses include site selection (see Chapter 1, "Marketing, Advertising, and Sales"), property management, demographic analysis, and economic development, but the list of applications grows frequently. The foundation for every real estate project is location, or more precisely *geographic* location, where location can be defined as a vacant lot, gas station, restaurant, condemned building, high rise apartment, 70-story hospital, or billboard. The geographic extent of the project can cover 1,000 sq ft or 3,000 sq mi. In short, if an object can be located by an address, within a city, using section/township/range information, or with latitude and longitude coordinates, it can be managed with a mapping tool.

Data: The Key to GIS

More than 80% of corporate business information is estimated to have a geographic component. Such components can include addresses, ZIP Codes, cities, or even latitude and longitude coordinates. This fact alone emphasizes the inherent suitability of business data for mapping technologies. However, the adage "garbage in, garbage out" holds true particularly for GIS: the better the data entering the system, the more valuable the answers you will retrieve.

Three sources of data can be imported into a MapInfo system: client data, third party data, and MapInfo data partner products. Each type is described below.

Client Data

As the client, you may have a substantial amount of the data you require in house, although you may not know it. Information on customers, competitors, historical finances, and even hand-drawn maps of sales territories can be imported into MapInfo. The software can read a variety of data formats, including ASCII, Microsoft Excel and Access, Lotus 1-2-3, and Dbase files. With open database connectivity, remote data from larger database systems can also be imported from Oracle, Sybase, Informix, and SQL Server. If your company maintains these data files in an electronic format, they can be imported into MapInfo as distinct layers.

Third Party Data

Many companies have purchased market and competitor data from consulting firms, data vendors, and other sources over the course of several years. If the third party data files are also in one of the electronic formats described above, they can be imported as well. Many data vendors have specialized databases built specifically for GIS applications because some databases have driven the need for GIS (in contrast to situations where the need for GIS has driven data collection).

Determining that no data exist for a particular application is not a dead end either, because it can become an opportunity to collect data that may have market value for other GIS users.

MapInfo Data Partner Products

Recognizing the value of high quality data, MapInfo Corporation has become a significant source for virtually every type of data. From street files to boundary files and specialized databases, the company either offers data or partners with data vendors to ensure that end users have abundant choices. After all, GIS may come and go but the market for useful, accurate, and timely information is likely to persist and expand.

Finally, as part of its international offerings, MapInfo Corporation offers coverages or databases for a number of foreign countries. Access the company's Web site at *www.mapinfo.com* for more information.

Site Selection

Locating an optimal site can be problematic and mysterious, particularly in developed and mature markets. In GIS, the right tools can often spell the difference between failure and success. Mapping technology has transformed site selection into a scientific methodology, whether the site is for a new retail store or aquatic farm. Most site selection systems replicate the manual and mechanical process but with the added benefits of speed, lower cost, and formalized methodology.

Under the manual process, a real estate manager would drive to a potential site, walk the property, view the landscape, and observe traffic patterns to develop a "gut feeling" or instinct regarding the location. She would drive major streets in the vicinity, investigate the competition, and take in the commercial and residential makeup of the area. At the office, after gathering additional information, the real estate manager would stick a pushpin in the city wall map to signify the selection.

With mapping technology, the real estate manager turns on her computer and accesses MapInfo's real estate application. Using MapMarker (the geocoding function), she types in the cross streets of the potential site. Accessing the data layers, she overlays current street files, traffic counts, block group demographics, and competition files, in order to view relevant relationships. With this overlaid information she can identify the flow, density, and average annual daily traffic volume on primary and secondary streets. She analyzes the growth and trend of population using 1980, 1990, current year and five-year projected population counts by block groups. Age, income, and occupation data are quickly mapped to profile the potential customer base in the trade area using 1, 3, and 5 mi rings. Competitor locations are identified relative to customers and traffic flows. Daytime population counts by block group are mapped in colorful themes to identify the concentration and amount of potential daytime customers. Once a site has passed the initial review using MapInfo, the real estate manager may travel to the site.

While instinct certainly plays a role, in the current business environment real estate managers are likely to be responsible for many states and locations. Without a tool to screen hundreds of potential new sites and target the strongest candidates, real estate managers fight a losing battle. For more information on this application, see the "Visimark" case study in Part Two.

Real Estate Management

Property management is a fundamental application for business geographics. Locating and extracting site, demographic, trade area, and market information is straightforward and easy to understand. The layering function enhances information and allows users to quickly access data. Among other applications, real estate companies can easily generate population/demographic maps and reports for potential property lessors.

In addition, gap analysis can determine "missing" offerings in certain market areas. For example, it would benefit both the landlord and tenant of a strip mall to fill a vacancy with the most appropriate business. If 10 dry cleaners are located within 3 mi of the mall, another dry cleaner may be unsuccessful. However, using mapping tools, the landlord can determine the competitive mix in 1, 3, and 5 mi rings, analyze the area's demographics, and determine the services and products that best fill the gap. Mapping refines gap analysis and provides a tremendous presentation vehicle for selling the location to potential tenants.

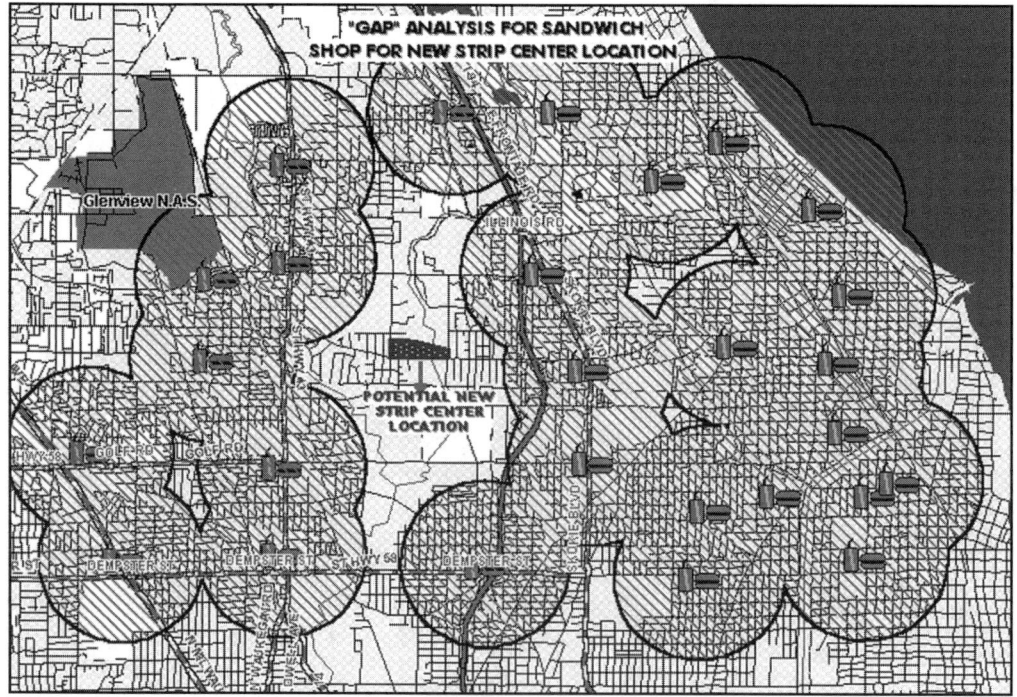

Gap analysis map of a retail site for a new strip mall.

Improving the Home Search

GIS is revolutionizing the home search process. Time constraints, the desire for detailed consumer information, and the need to make decisions quickly have all spurred the use of GIS in home sales. Home buyers can electronically search for new homes in realty offices or over the Internet. Instead of driving extensively to inspect potential homes, customers outline their requirements (e.g., price, square footage, neighborhood) and amenities (e.g., surrounding schools, floor coverings, shopping, interstate access) at the realty office. The realtor queries a database, and a map is generated with homes meeting the criteria. Photographs and videos show home interiors and exteriors. The map illustrates nearby schools, shopping, and driving routes. Digital home buying services offer a worthwhile alternative for customers and realtors alike, a win-win situation.

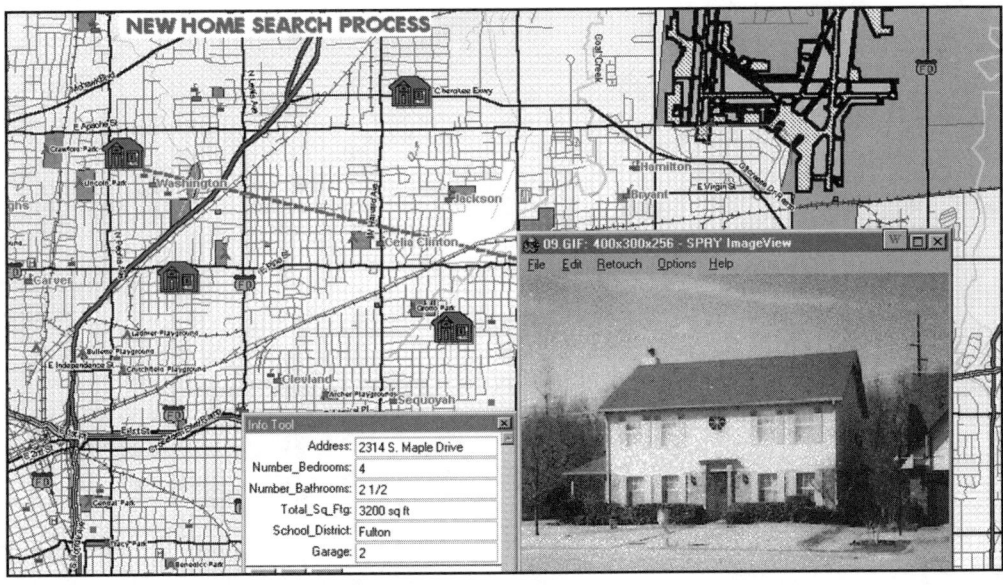

Home buying service using mapping technology.

PART TWO

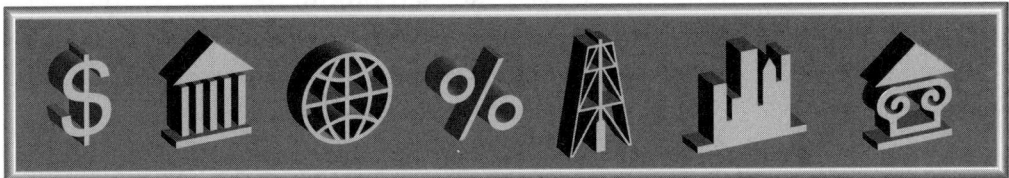

CASE STUDIES

This section presents a wide variety of case studies involving real world applications of MapInfo Professional and other software and tools in the MapInfo product line. As is often the case with MapInfo, while the desktop mapping software may have been implemented to meet a specific objective (e.g., target marketing), in some companies its use has been significantly expanded (e.g., facilities management, evaluating hypothetical scenarios, and making presentations). Consequently, the primary uses of MapInfo within each case study are listed at the beginning of each discussion.

The authors gratefully acknowledge the efforts of the individuals responsible for each case study; contributors are listed after each case study's introduction on the following pages. Additional contact information for contributors can be found in Appendix D.

- **Price changes in retail petroleum markets.** A prototype was developed by the book authors to determine how customers respond to price increases of unleaded premium gasoline in order to maximize profits for a national petroleum retailer. (IntelleVue)

- **Determining the customer base for a national car rental franchise.** Thrifty Rent-A-Car uses MapInfo to analyze where its customers originate and target future marketing efforts more effectively. (Rick McDowell and Sandy Carter of Thrifty)

- **Advertising and target marketing in regional publications.** Based in Australia, Rural Press merges the mapping and analysis capabilities of MapInfo with CDATA, electronic census data developed by the Australian Bureau of Statistics, to attract advertisers and supports strategic planning. (Kareena Kinnear and Alan Pont of Rural Press, and Craig Chevrier of MapInfo Corporation)

- **Siting selection.** The Retail Analysis Department of Canada Post Corporation combines modeling techniques, detailed retail data, and mapping tools to site new retail postal stores. (Paul Bernard, Peter Halpenny, Teresa MacKenzie, Martin Sarch, and David Slaughter of Canada Post Corporation)

- **Emergency response.** Located in Pittsburgh,

Pennsylvania, the Emergency Medical Service Institute (EMSI) uses MapInfo and software developed by On Target Mapping to enforce regulatory guidelines governing the operations of 200 ambulance services within its jurisdiction. (Christopher Price of EMSI and Debra Magee of On Target Mapping)

- **Telecommunications**. The largest wireless messaging company in the United States, PageNet uses MapInfo solutions developed by Public (PSA) to conduct target marketing, access real time information on towers, and optimize coverage among 9,500 transmitters. (Mary Coffee and Rick Peters of PSA)

- **Oil and gas well mapping**. Louis Dreyfus Natural Gas employs MapInfo to map the locations of oil and natural gas wells, manage leases, and plan future drilling activities. (Stan Nickel of LDNG)

- **Real estate**. Visimark integrates MapInfo, diverse databases, and electronic images in its Internet-enabled application for commercial real estate brokers and agents. (Roberta Nelson-Walker of Visimark)

CASE STUDY
RETAIL • MARKETING

PRICE CHANGES IN RETAIL PETROLEUM MARKETS

Accurately predicting how customers react to price changes is a goal of all gasoline retailers because such information can enable them to control price and profitability on a store by store basis. In the retail petroleum industry, the gasoline with the greatest profit margin is unleaded premium, the most expensive grade. However, while profit margins on unleaded premium are dramatically higher than other grades, unleaded premium only accounts for 10 to 15% of gasoline sales in most markets.

Knowing in advance how customers react at each store enables retailers to maximize profits and maintain sales—with profound implications for the bottom line. For example, a company with 1,000 stations selling 80,000 gallons of gasoline per month would earn nearly $3,000,000 more each year if the profit margin on unleaded premium rose an average of 2 cents per gallon (assuming that unleaded premium accounts for 15% of sales).

Prototype Mission

IntelleVue (formerly J. Davis & Associates, Inc.), was contracted by an international petroleum retail marketer to build an application prototype using MapInfo. As a first step toward identifying customer price reaction on a store by store basis, IntelleVue was asked to track customer purchasing trends in relation to changing market conditions.

The project involved gathering oil company credit card purchases over time from sample sites that reflected stable market conditions (except with regard to price). The prototype tracked customer purchases by credit card number, capturing data on date of purchase, store, gasoline grade and gallons purchased, and the dollar amount of each transaction. Customer anonymity was maintained because only oil company credit card numbers and ZIP+4 ZIP Codes were used. Bank credit card and cash purchases for each store were summarized as total figures. The geographic locations of these customers were assumed to reflect the sample credit card customer database.

Before MapInfo was used, sample stores, timing methods, and loyal customers had to be identified.

- **Sample stores**. Three locations representing low, middle, and high sales under stable market and competitive conditions were selected. Locations within several miles of one another were chosen on the assumption that they shared customers.

- **Sample data timing**. Twelve months of credit card data were obtained during a period where no sample sites experienced operational or facility changes.

- **"Loyal" customers**. To track customer reactions most effectively, a sample of customers who purchased at certain stores on a regular basis was identified. Loyal customers were defined as individuals who purchased a minimum of two times per month using an oil company credit card.

Tracking Customer Purchases

Changes in customer purchases were classified according to four categories. The prototype assumed unleaded premium customers would react to price increases in one of four ways.

- **No change**: The increase did not alter purchasing behavior.

- **Buy down**: Customers purchased a lower grade of gasoline at the same location.

- **Company crossover**: Customers purchased gasoline from other locations within the same retailer (i.e., they remained brand loyal).

- **Lost customers**: Customers ceased purchasing gasoline using the company's credit card (no additional purchases were recorded). It was assumed the customers were lost to competitors.

NOTE: The results identified in the prototype were not intended to be exhaustive. Some customers may have switched for reasons other than price.

MapInfo Screens and Application Programming

MapInfo was chosen by the retailer for the project. Fortunately, client data were stored in Microsoft Excel spreadsheets and Microsoft Access data tables—formats that could be easily imported into MapInfo. Necessary mapping functions were programmed in MapBasic, the MapInfo developer's environment, as a single *.mbx* application that altered MapInfo menu items.

The following menu items and dialog boxes were created. Main menu bar selections were modified for user friendliness. All standard MapInfo menu items were maintained but switched to a vertical format under a single menu item. This provided full MapInfo functionality while freeing space for custom menu features and a simplified menu bar.

MapInfo Screens and Application Programming

Main menu item	Function
MapInfo	All basic MapInfo menu functions
Site-Market Selection	Select a market or a site
Quit	Exit

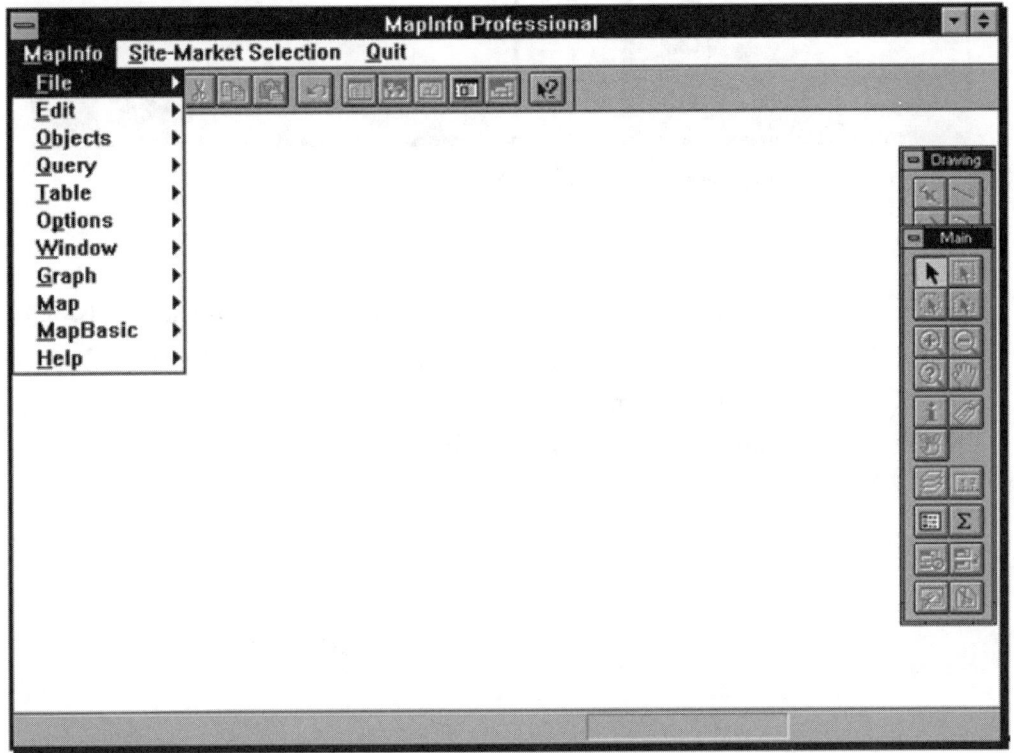

Main prototype window.

Choosing Site-Market Selection from the main menu bar displayed the following custom dialog boxes developed in MapBasic. The boxes enabled the user to select the appropriate market (Select a Market), and then a sample location and a zoom setting for the map (Select a Site).

 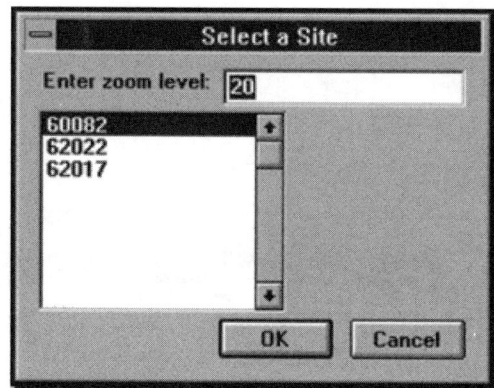

Select a Market (left) and Select a Site (right) dialog boxes.

Site Map

When a user chose a site, a map appeared with the site, streets, and census block groups visible. (Other client sites in the 20 mi area were also visible.) A Pricing Analysis menu was also displayed.

MapInfo Screens and Application Programming   149

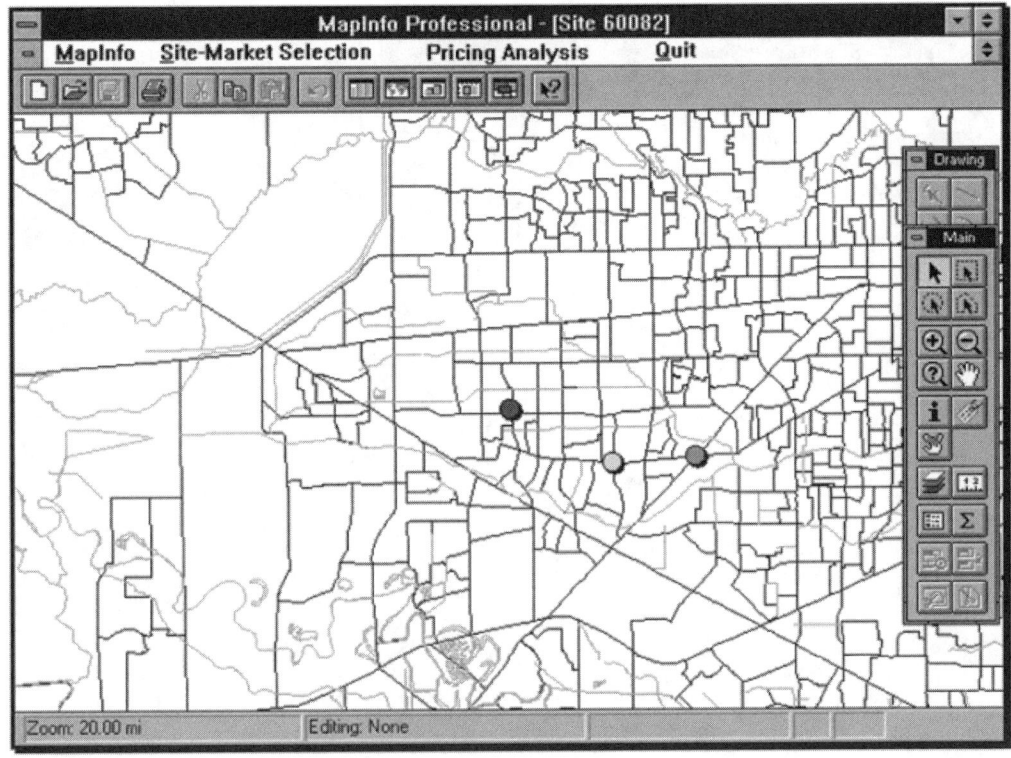

Site map.

Pricing Analysis Menu

The Pricing Analysis menu was developed to support several types of data analysis. When invoked, it generated a map displaying customers as points to identify proximity and trade areas, and thematically shaded census block areas by various values (e.g., customer density, gallons sold, and total dollars sold). In addition, the analysis graphed the four customer reactions previously outlined. The percentage of sales by grade and type of transaction (company credit card, bank credit card, and cash) for individual locations was also displayed.

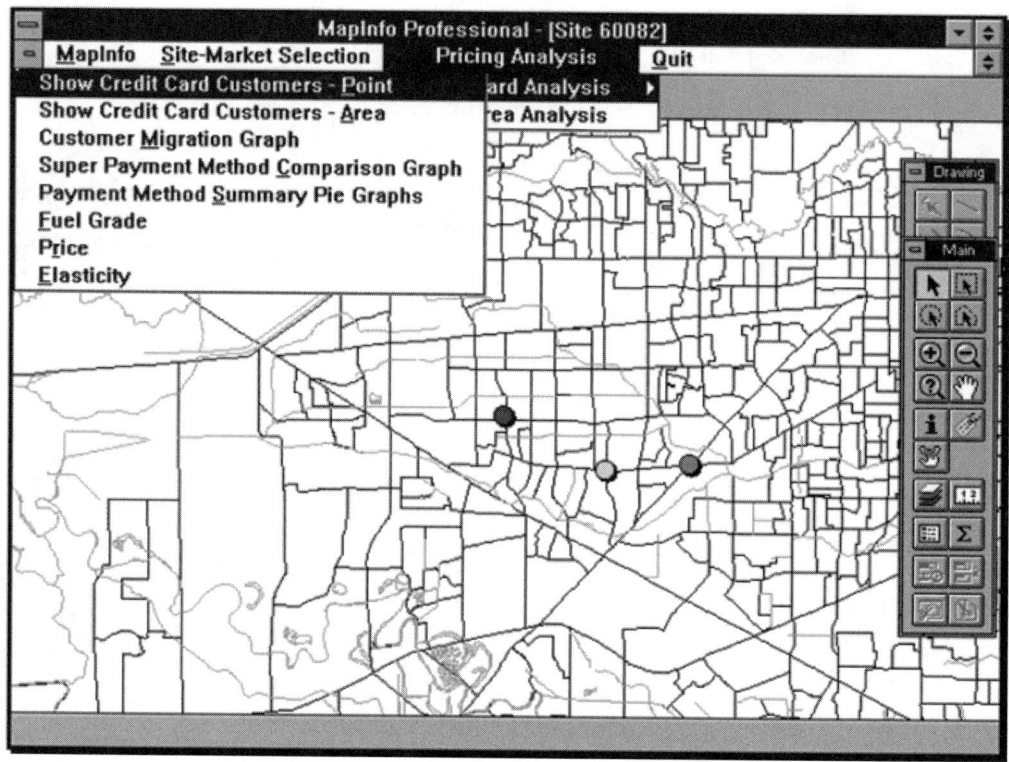

Pricing Analysis menu.

A short description of each item under the Pricing Analysis | Credit Card Analysis menu follows.

Item	Function
Show Credit Card Customers - Point	Customers appear on map as stars
Show Credit Card Customers - Area	Census blocks shaded by concentration of customers
Customer Migration Graph	Bar graph detailing volume loss due to • Same store, lower grade • Another client store • Competition (lost volume)

Item	Function
Super Payment Method Comparison Graph	Pie charts showing payment method for unleaded premium
Payment Method Summary Pie Graphs	Purchases by payment method for all grades
Fuel Grade	Function unavailable in prototype
Price	Function unavailable in prototype
Elasticity	Function unavailable in prototype

Credit Card Customer Point Map

The Show Credit Card Customers - Point menu option generated a dialog box allowing the user to display credit card customers for the selected site, other client sites, and common customers among various client locations. (Common customers were defined as customers who purchased products at two client sites in the same time period.) If the user selected Other Site, a subsequent dialog box listed valid sites.

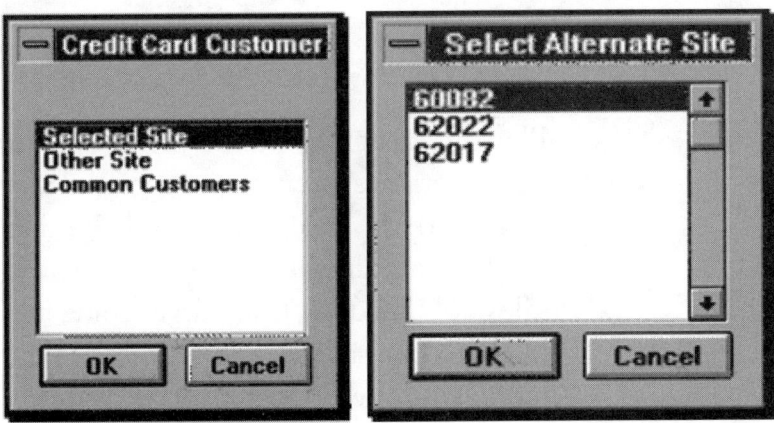

Dialog boxes accessed when the Show Credit Card Customers - Point menu option was chosen.

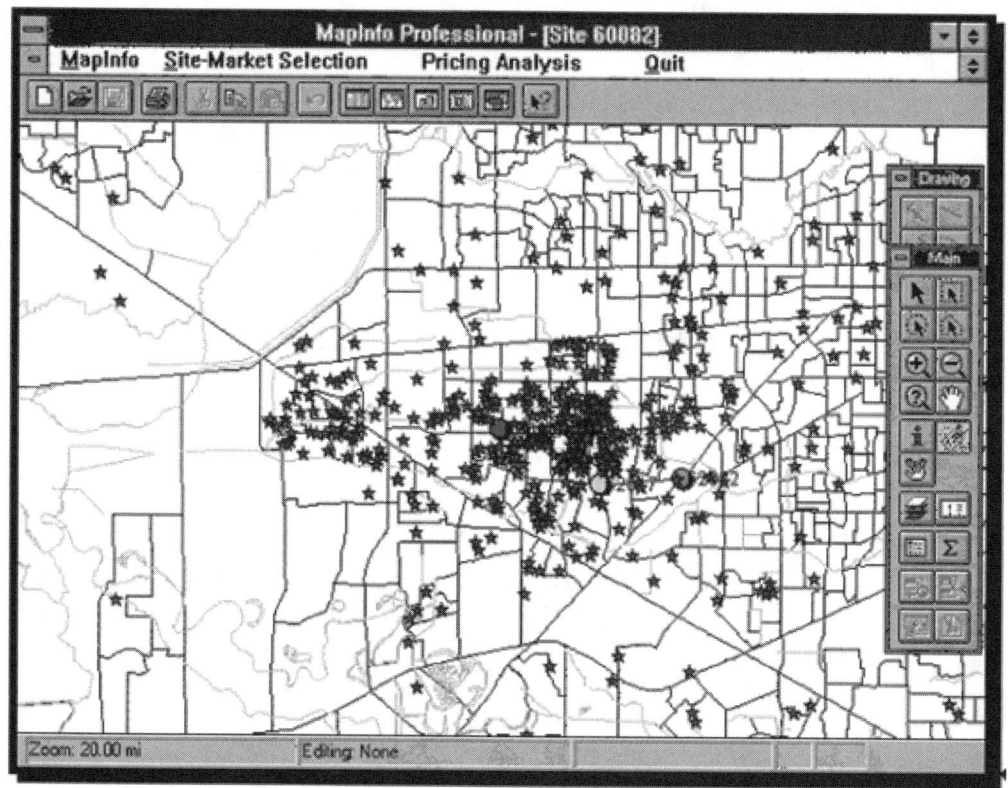

Sample map of plotted customer points.

Plotting customer points on a map highlighted several characteristics of customer patterns for a particular store, such as trade area density and the distance from which the location drew customers. Typically, sites with higher sales volume drew from larger trade areas and were more likely to lose sales as a result of price increases. Sites with more customers located at a distance generally lost a greater volume than sites that drew from a smaller area. While such analyses were possible, the overall intention of the customer point maps was to quickly show where customers were located relative to a given site.

Common Customers

Plotting credit card customers who had purchased gasoline at two locations during the same time period generated a map similar to the following illustration.

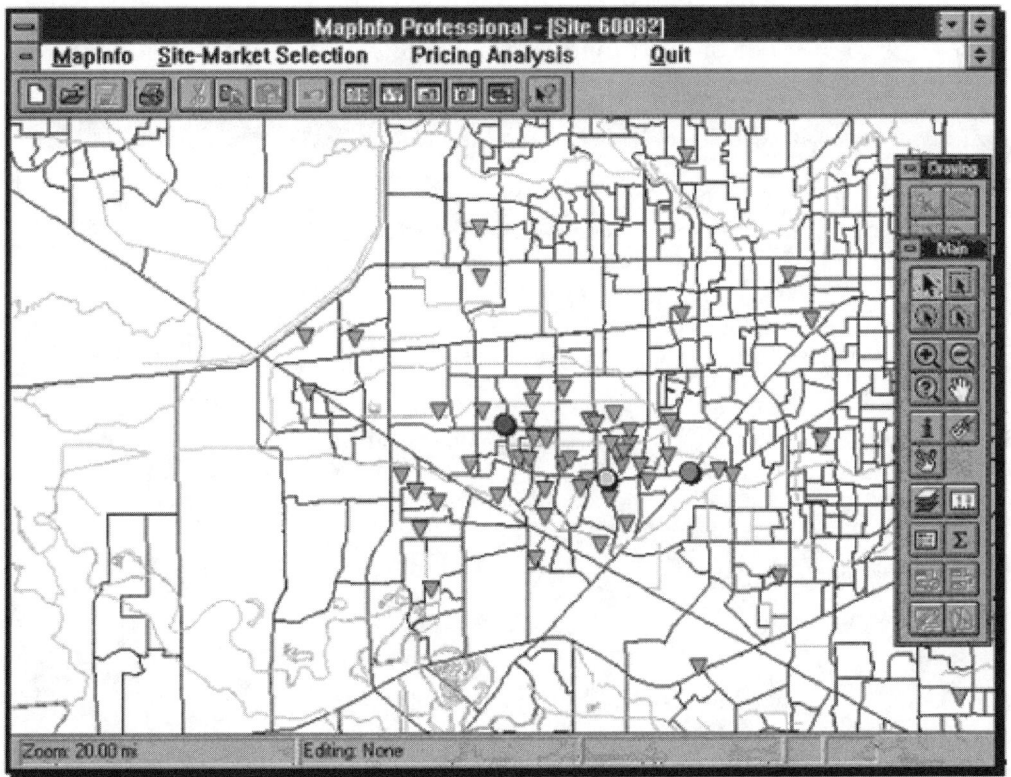

Sample map showing customers common to two locations.

◦ **NOTE:** *The number of current customers who purchase at two locations may be an indicator of brand* *loyal behavior. Such potential crossover customers are important because they are likely to remain customers in spite of price changes.*

Credit Card Customer Point and Area Maps

If the user chose the Show Credit Card Customers - Area menu option, thematically shaded census block groups associated with the customer data table were displayed. The thematic shading showed three variables: total number of customers, total gallons sold, and total dollars sold.

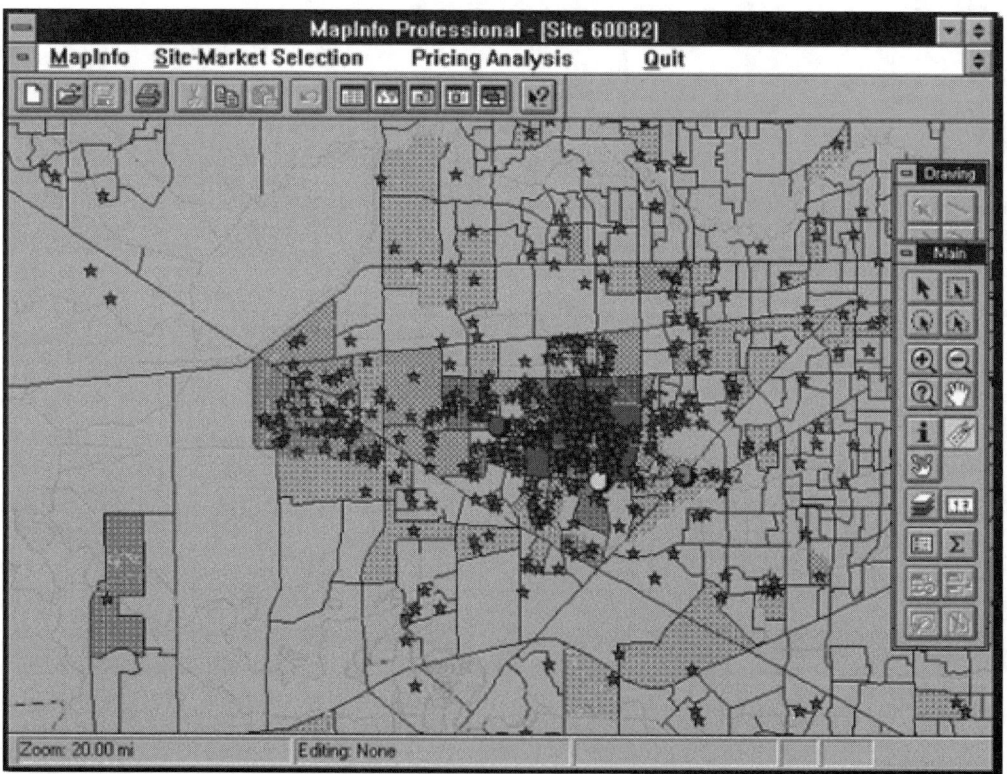

Sample customer area map by total number of customers, total gallons sold, and total dollars sold. (The distinctions were readily apparent when the map was generated in color.)

Area maps identify density and other characteristics of customer purchases surrounding the client's site (thematically shaded by number of customers in each census block group). Typically, the closer the concentration of purchases (by number of customers, gallons, or dollars) to a site, the more local the customer support. On the other hand, a more dispersed sample generally indicates more customers are in contact with competitive facilities and they are more likely to respond to market changes such as price fluctuations, new facilities, road changes, and so forth.

Parameter Choices

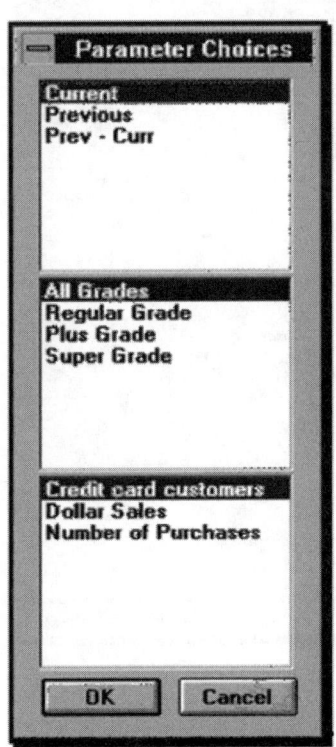

Parameter Choices dialog box.

For each map previously described, the user could choose from three areas to display data in different formats. These options allowed the user to "drill down" to specific data and generate maps to help explain customer behavior at a site. Available parameters (which appear in the image at left) follow

- **Time frame**: Current, previous, or the difference between current and previous time frames.

- **Gasoline grade**: All grades, unleaded regular, unleaded mid-grade, or unleaded premium.

- **Data category**: Actual credit card customers, dollar sales, or number of purchases.

The parameters allowed the user to sort and sift data in multiple ways to enhance feature analysis. For example, the client could focus on dollar sales of unleaded regular gasoline for the current time frame.

Customer Migration

Customer migration was shown using a map and bar graph that illustrated volumetric change between two time periods.

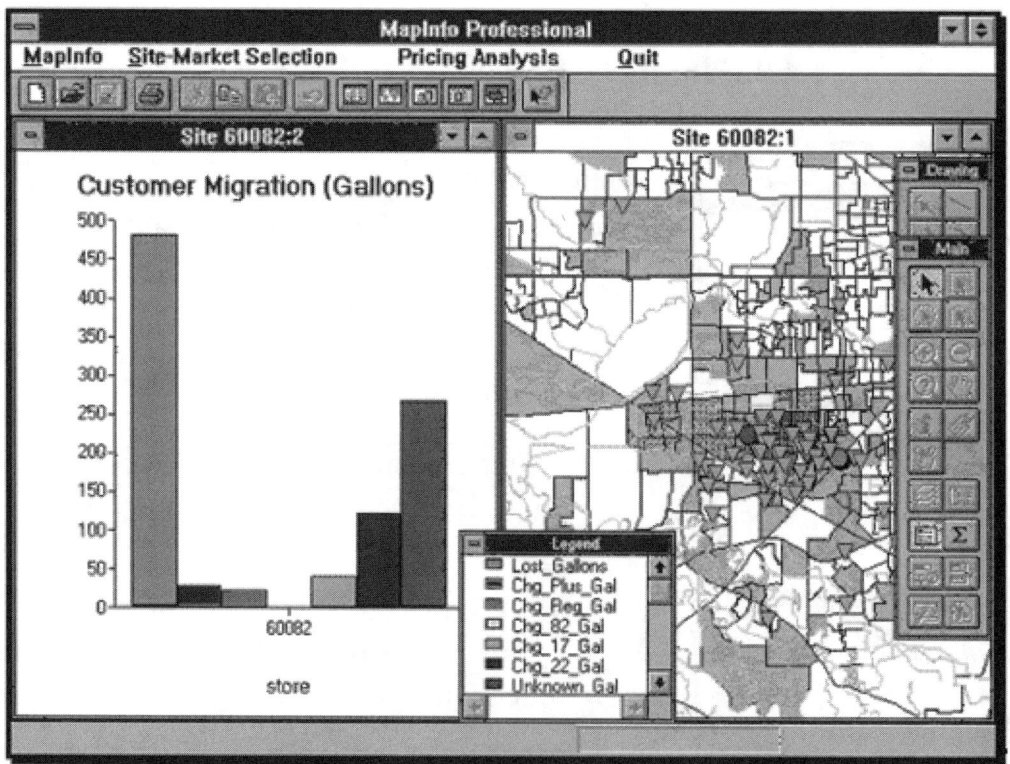

Sample customer migration map and graph.

The above graph highlights unleaded premium volume loss and where it changed within the selected time frame.

The bar graph shows the following information (from left to right).

- Total volume change (*Lost_Gallons* in the Legend)

- Premium gallons that switched to mid-grade gasoline (*Chg_Plus_Gal*)

- Premium gallons that switched to regular gasoline (*Chg_Reg_Gal*)

- Premium gallons that remained—no change (*Chg_82_Gal*)

- Premium gallons taken up by other client sites (*Chg_17_Gal* and *Chg_22_Gal*)

- Premium gallons lost to unknown competitors (*Unknown_Gal*)

> **NOTE:** *The pricing history of all sites was not included in the prototype. The next step in the analysis would logically be to include a line graph representing the street price of the selected location as well as all other affected sites. The map displayed could use any combination of parameters discussed previously.*

Analysis

With this information, the project sought to identify changes in purchasing habits within the selected time frame. The chart provided the basis for estimating the impact of a price change and the volume associated with customers who purchased a lower grade of gasoline at the same location. It further identified customers who switched to another site in the same chain or who switched gasoline retailers altogether. The effects of each change follow.

- Buy down customers are less likely to shop around; they are valuable, loyal customers.

- Crossover customers react to price but remain brand loyal.

- Lost customers are price conscious shoppers not loyal to brand or store.

The ability to analyze customers according to such categories is a strong starting point toward understanding each site's customer profile—which in turn offers insight regarding how to label stores for price analysis.

Theoretically, stores with high concentrations of buy down and some crossover customers would likely absorb price increases in unleaded premium gasoline, allowing them to maintain sales volume while boosting profit margins. On the other end, stores with high concentrations of lost customers would typically sacrifice sales volume with price increases. The trick is balancing volume loss and additional profits. If the increased profit from the higher price more than offsets the lost volume, the site may have a winning formula. The ability to develop a customer profile for each store represents a significant step toward providing the knowledge necessary to accurately assess potential customer reactions to price increases and allow for store to store price management.

Premium Payment Method Comparison

The next image shows premium gasoline purchases for the selected time frame and purchases broken down by payment method (company credit card, bank credit card, or cash purchase).

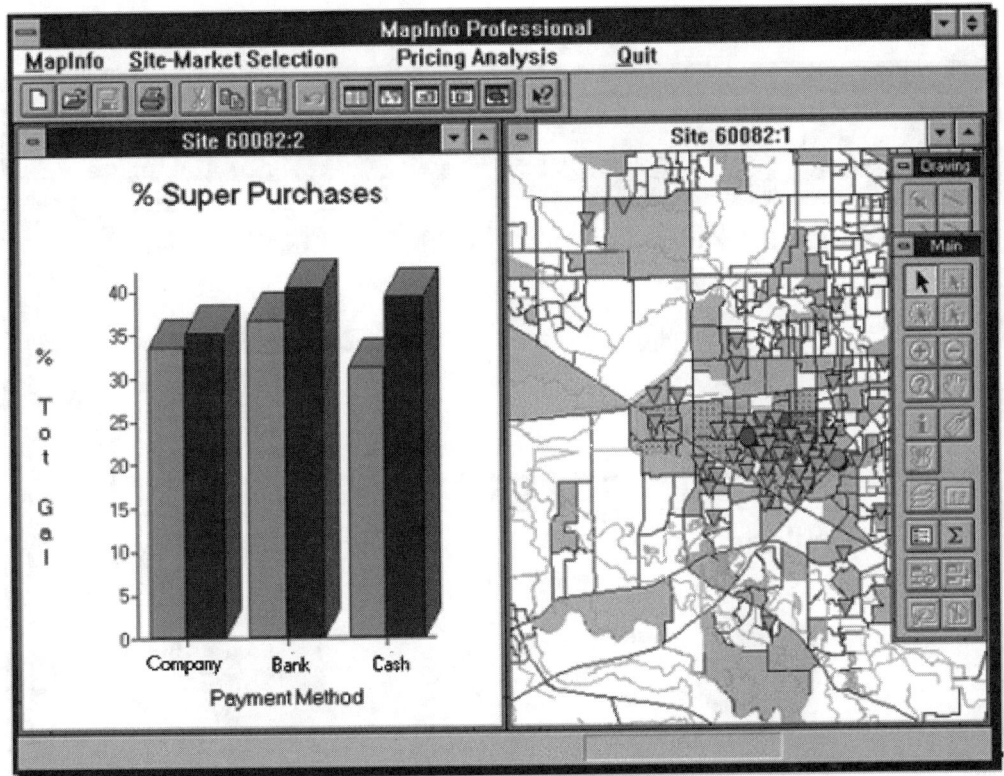

Sample premium payment method comparison.

This approach analyzed purchase patterns by transaction. (Recall that only client credit card transactions were analyzed; IntelleVue assumed that other credit card customers followed a similar pattern.)

Payment Method Comparison

If the user selected the Payment Method Summary Pie Graphs menu option, a screen showing four pie graphs was displayed: the breakdown of premium gasoline purchases by payment type, and three graphs depicting how sales of different grades are divided by payment method.

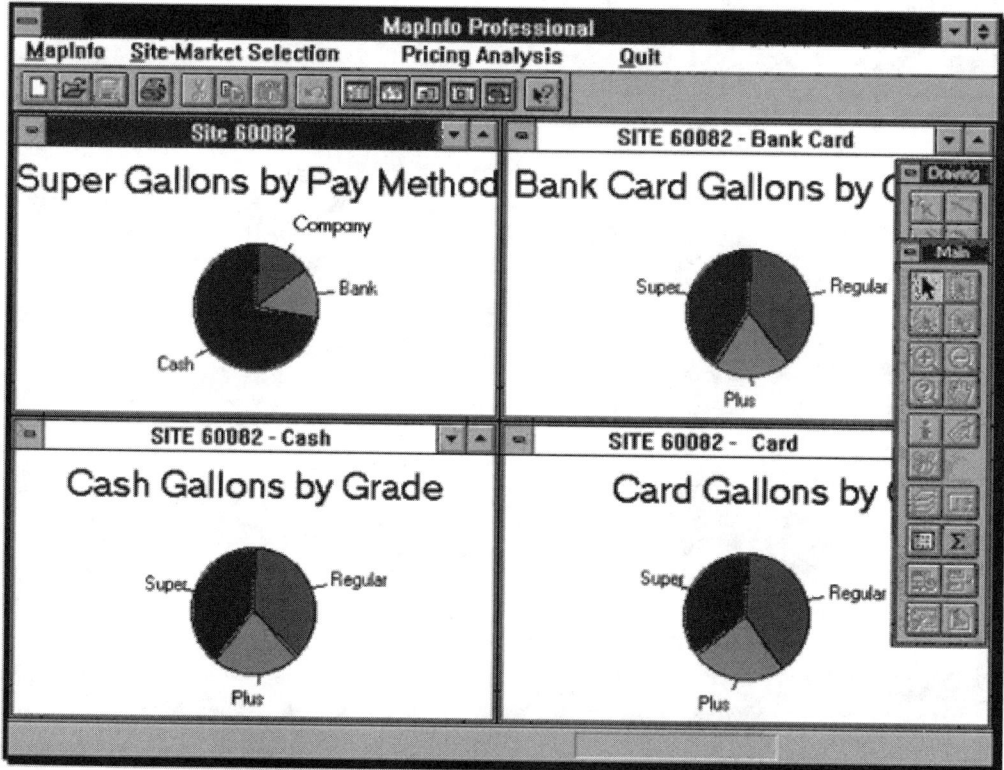

Sample analysis of premium gallons sold by payment method (top left graph), and how sales of different grades are divided by payment method (remaining graphs).

Conclusion

Understanding customer behavior in relation to price can be viewed as simple (i.e., price goes up, sales go down) or extremely complex. Most marketers have expressed both opinions and at heart believe both statements to be true. The quest to find the answer to this fundamental marketing question will continue, and the individual who unlocks the mystery—and can accurately and consistently predict customer reactions—will find a pot of gold waiting at the end of the price rainbow.

Acknowledgment

Thanks to Bambi Murphy of Tulsa, Oklahoma, the GIS developer for the project.

CASE STUDY
MARKETING

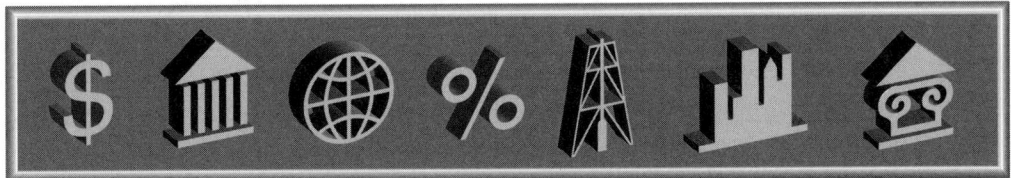

THRIFTY RENT-A-CAR

Thrifty Rent-A-Car System, Inc., recently introduced MapInfo software into its marketing and strategic planning processes with great success. Thrifty currently uses MapInfo to access and analyze data for strategic and marketing purposes. Mapping customer data allows the company to uncover relationships, patterns, and trends that spreadsheets and database reports cannot. While the use of MapInfo at Thrifty remains new, the company's director of U.S. Licensee Operations, Rick McDowell, impressed many of the company's top franchise owners with a demonstration of how even the most profitable locations could boost their bottom lines with business mapping.

Know Your Customers

In addition to MapInfo software, high quality data have been critical for the Thrifty application to succeed. Thrifty owners rely on customer information gathered electronically from 425 locations in the United States and Canada. Currently, 92 percent of all active locations in the United States electronically transmit data to the company's headquarters in Tulsa, Oklahoma. Timely access to customer information allows Thrifty to pay travel agent commissions more efficiently, improves the relevancy of nationwide statistics about Thrifty's customers, and supports the auditing of sales reports.

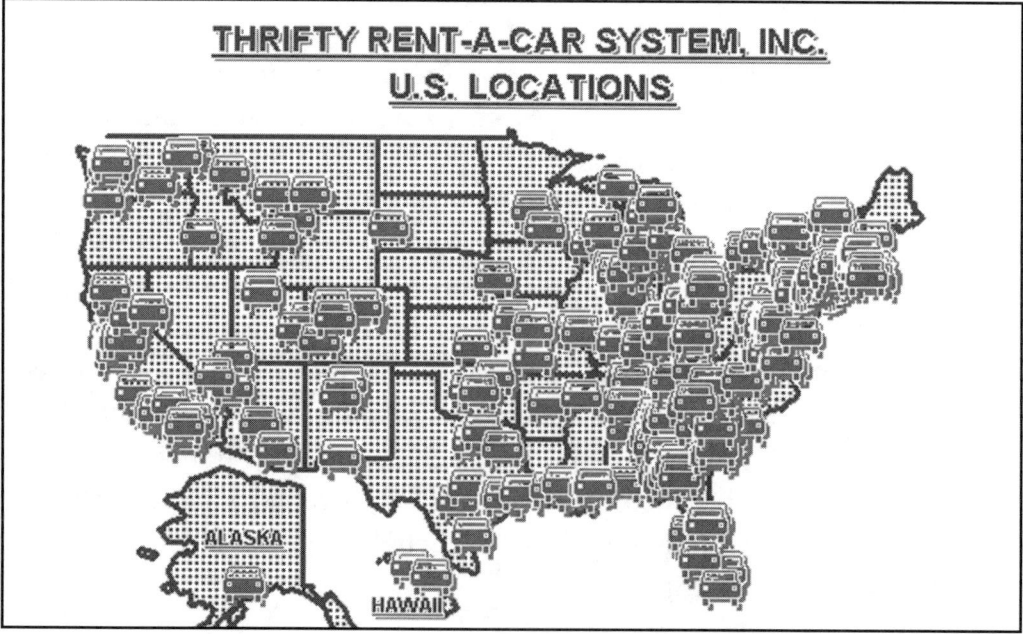

Thrifty's U.S. locations that gather and electronically transmit data to the Tulsa headquarters.

Analysis of closed contract data revealed two important facts even before they were imported into MapInfo for analysis. Thrifty discovered that 80 percent of renters are male, which encouraged the company to focus more advertising on the female population, such as the sponsorship of the U.S. National Figure Skating Championships. In addition, the company learned that its over age 55 business segment was not keeping pace with the national average. Consequently, Thrifty is increasing marketing efforts toward senior citizens.

Generating Maps

Rental contract data collected in Tulsa are imported into MapInfo to generate maps showing where customers reside. Specifically, the MapInfo team concentrated on the company's largest volume stores. McDowell and colleagues Sandy Carter (Section Manager, Statistical Resources) and Sharon Waner (Coordinator, Statistical Resources) plotted 357,285 ZIP Codes of 1.2 million customers and generated 225 customer origination maps.

Carter, who has since assumed mapping responsibilities at Thrifty, noted that MapInfo enables the company to view customer points of origination at global or extremely detailed views. In addition, the software plots data on a thematic, or gradient, scale. Initially, three customer origination maps were generated for each owner of a licensed territory using customer data from the 1996 calendar year. They were provided a U.S. map, state map, and local or city map.

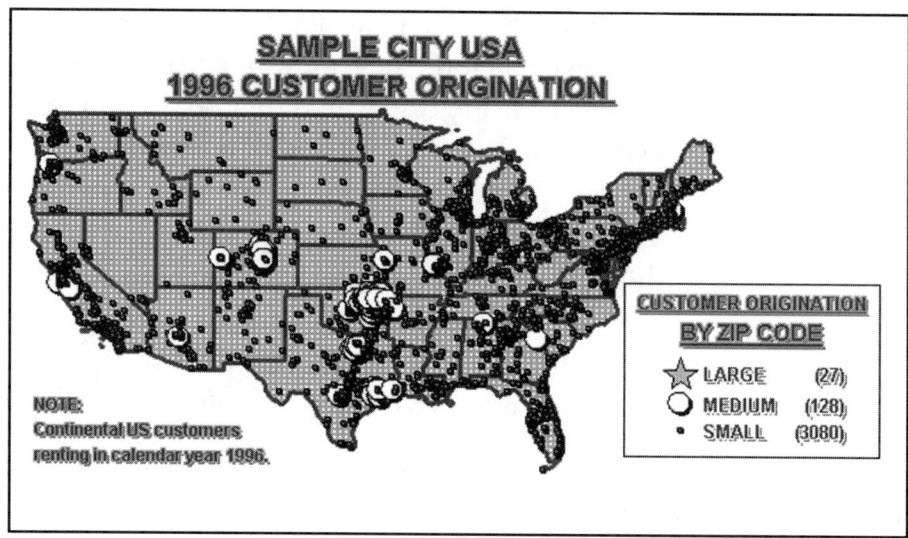

U.S. customer origination map.

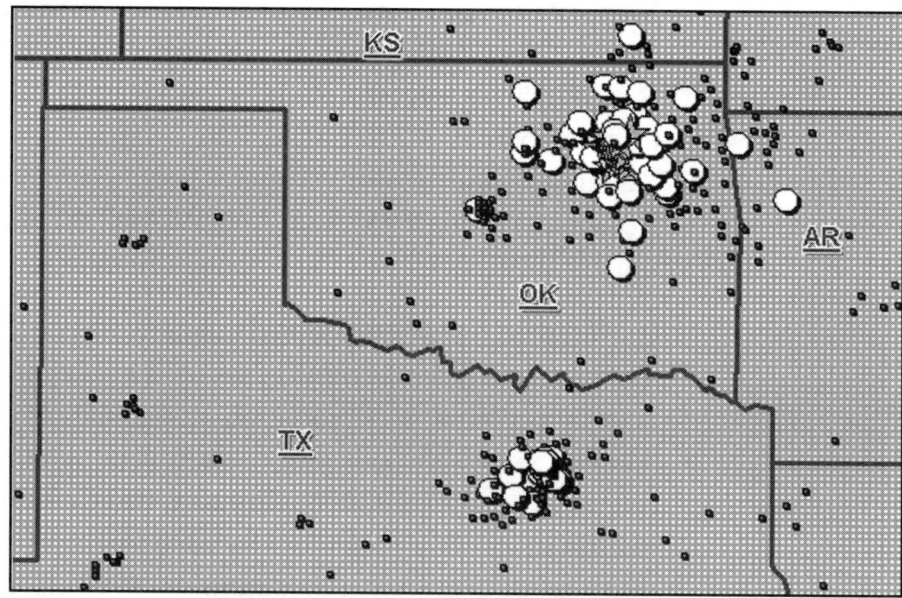

State customer origination map.

Emerging Trends

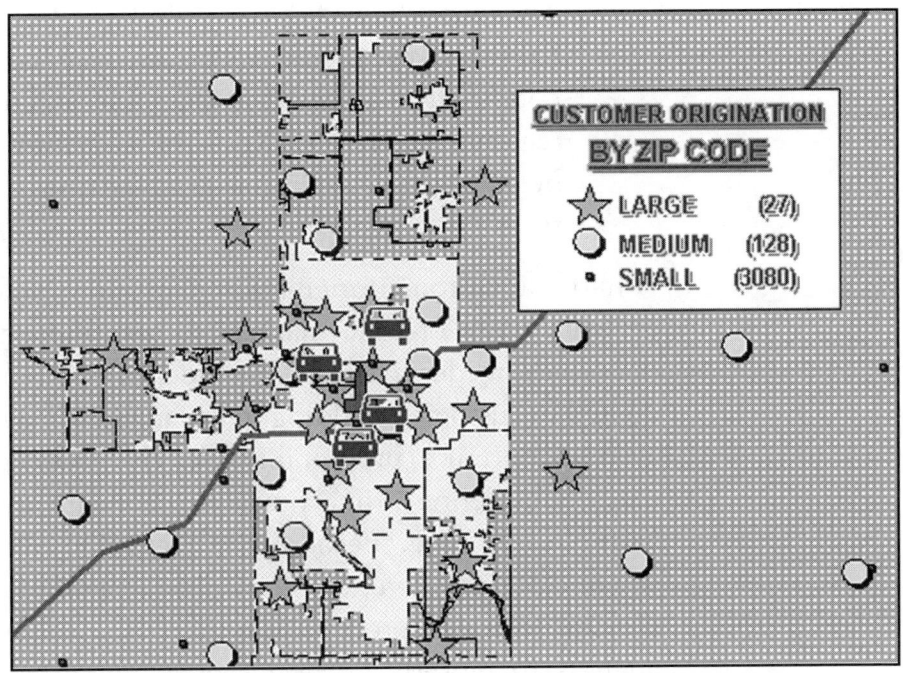

City customer origination map.

Emerging Trends

After mapping the Dirty Dozen data, the following national trends emerged.

- Three of the 25 owners' locations were characterized by incredible diversity in customer origination points. More than 10,000 ZIP Codes were plotted on each U.S. map for Orlando, Houston, and Atlanta. Others had limited regional draw.

- Maps for all airport cities showed that a large number of customers originate in the Eastern seaboard.

- The maps showed that certain key cities "feed" the rest of the country. The discovery could potentially trigger a shift from national to regional advertising. Knowing such information helps Thrifty owners refine advertising expenditures.

A national customer origination map allows an owner to evaluate point-of-scale markets to determine the most important feeder cities. Once identified, rental rates quoted to customers residing in key markets may be managed to optimize profitability. At the state level, the maps help define primary versus secondary trade areas within each state. Because the maps indicate exactly where customers live, they can play a significant role in the site selection process. MapInfo also enables Thrifty analysts to plot the locations of competitors, airports, hotels, car and truck dealerships, and other features of interest. Finally, Thrifty is expanding its use of MapInfo to include valuable competitor information such as approximate sales volume and store spacing for trade area determinations.

Conclusion

Since the presentation to the owners' group was made, the information has already been put to use. Owners in Phoenix have opened a new store and have three other locations pending. In addition, they are planning a direct mail campaign targeting the highest volume ZIP Codes to announce the new locations. Owners in other cities

Conclusion

are considering expansions as well. Having specific data regarding customer points of origin takes much of the guesswork out of siting new locations.

Thrifty is also enthusiastic regarding the possibility of linking MapInfo with a global positioning system (GPS) to locate rental cars worldwide. There is also some discussion regarding making MapInfo data available over the company's intranet, allowing all owners instantaneous access to the information.

For now, MapInfo information is available to owners on demand. However, owners with requests may find themselves waiting in line. Since the presentation, the statistics department has been deluged with requests.

CASE STUDY
ADVERTISING • SALES

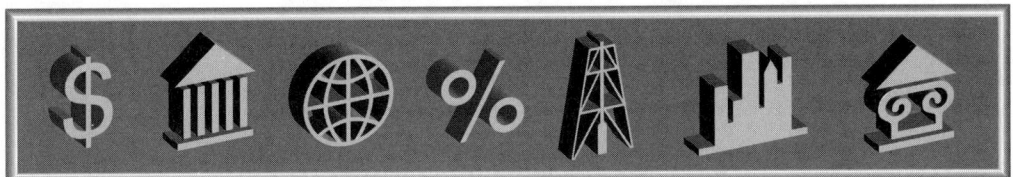

RURAL PRESS AND CDATA

Rural Press Limited specializes in agricultural, regional, and rural publications in Australia, New Zealand, and the United States. Its first publication, *The Land*, was launched in 1911 to foster stronger political and economic influence for farmers and grazers. The company's dedication to regional issues continues to guide its operations.

The cornerstone of agricultural publishing in Australia is the Rural Press network of state rural weekly newspapers with paid sales of more than 150,000 per week, and a readership of over 400,000. The following publications comprise the network.

- *North Queensland Register*
- *Queensland Country Life*
- *The Land* (New South Wales)
- *Stock and Land* (Victoria)
- *Stock Journal* (South Australia)
- *Farm Weekly* (Western Australia)

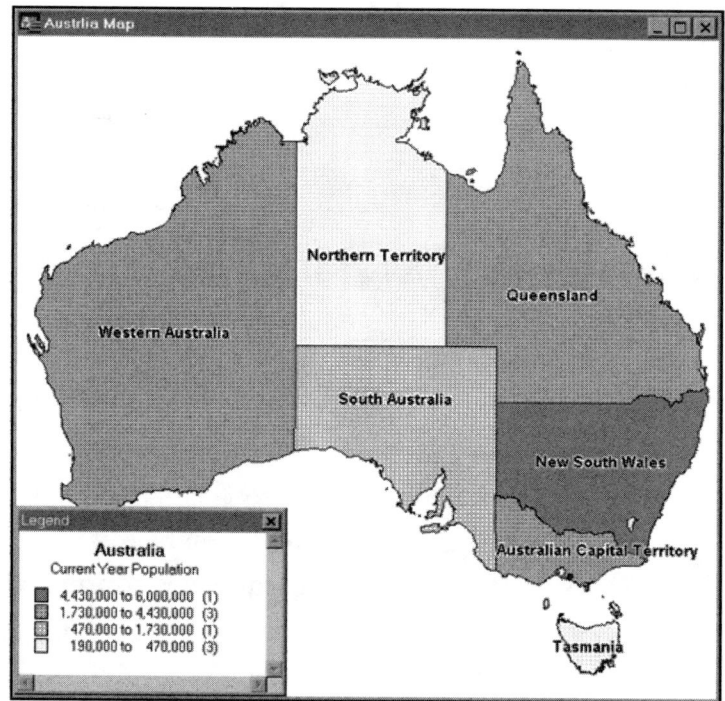

Australian states and territories by population.

In addition, the Agricultural Division of Rural Press publishes specialized national magazines aimed at highly defined market sectors.

Using CDATA

The Australian Bureau of Statistics (ABS), the official statistical organization for the commonwealth and state/territory governments, has developed a valued added data product to run with MapInfo software known as CDATA. The primary function of ABS is to collect information on a wide range of social and economic activities, compile statistics, and disseminate them to the government, the private sector, and the public. The current version of CDATA incorporates data on the following subject areas from the ABS Census of Population and Housing.

- Population, population projections, vital statistics, and migration
- Education, health, and welfare; justice; and other social issues
- National accounts, balance of payments, foreign investment, foreign trade, and public and private finance
- Labor force, employment conditions, prices, and household income and expenditure
- Agriculture, forestry, and fishing
- Research and development, manufacturing, energy, mining, retail and wholesale trade establishments, interstate trade, and tourist accommodations

Successes with MapInfo

With 140 publications in New South Wales, Rural Press struggled with the logistical nightmare of collecting and analyzing accurate statistical information on the small markets served by its papers. Acquiring information regarding publication performance in specific geographic regions was extremely time-consuming for staff in the regional division.

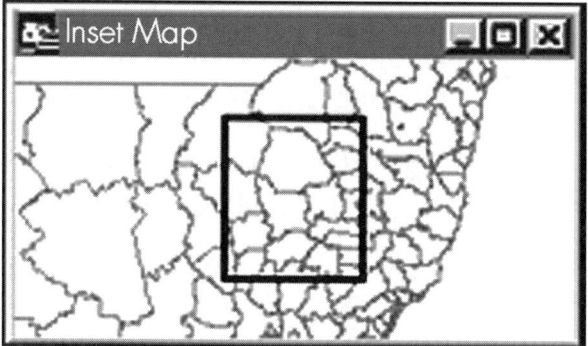

Geographic region of analysis.

When CDATA became available for use with MapInfo Professional, Rural Press could develop new statistics-driven sales and marketing efforts. Regional division staff use CDATA91 with MapInfo to produce a range of sales and advertising information without searching through ABS statistical books. MapInfo is used by the company to profile relevant communities and identify strengths and weaknesses in terms of age, income, disposable income, and population growth and decline. Such information is used for strategic purposes and to develop promotional material aimed at existing and potential advertisers.

Successes with MapInfo

The following map shows how Rural Press reviews population data using a map. First a thematic map depicts a population profile. Then a bar chart thematic map is added to display the breakdown of male and female versus total population.

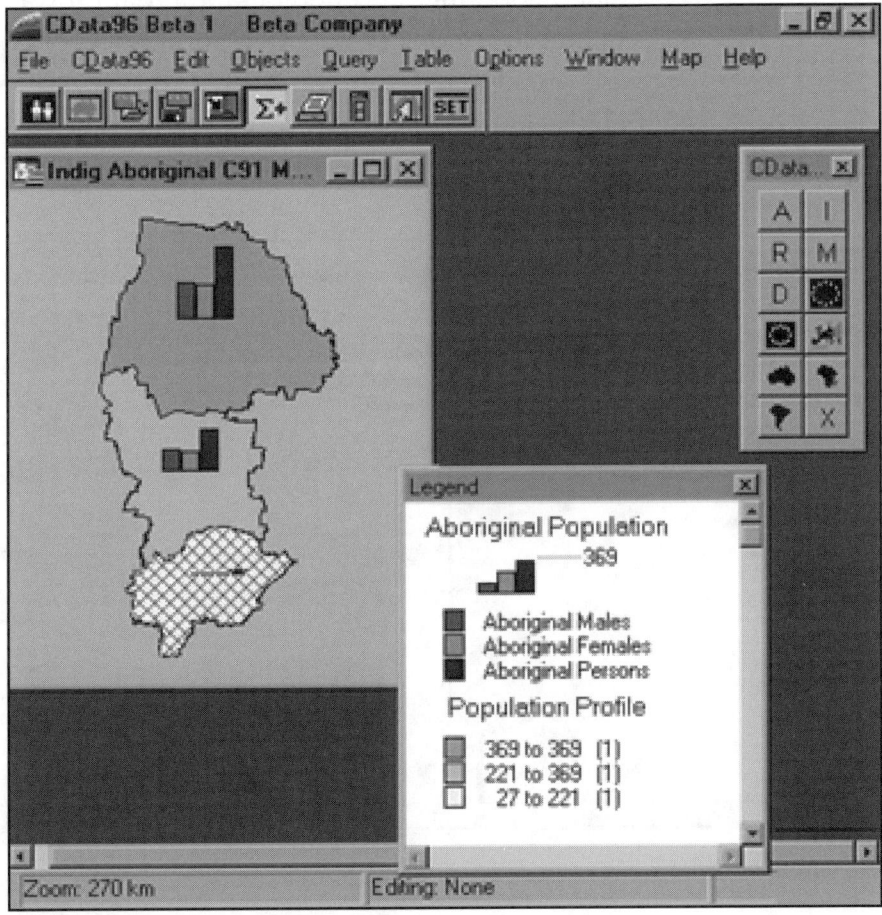

Aboriginal population analysis.

Rural Press uses the MapInfo statistics (or Data View) window for summarizing demographic information by region.

Taken together, CDATA and MapInfo are key sales and advertising tools that provide accurate statistics on social, economic, and demographic data for a variety of geographic regions (e.g., any combination of boundaries, rivers, roads, and unique land features). CDATA also includes an interface allowing users to quickly select the geography and data they wish to review.

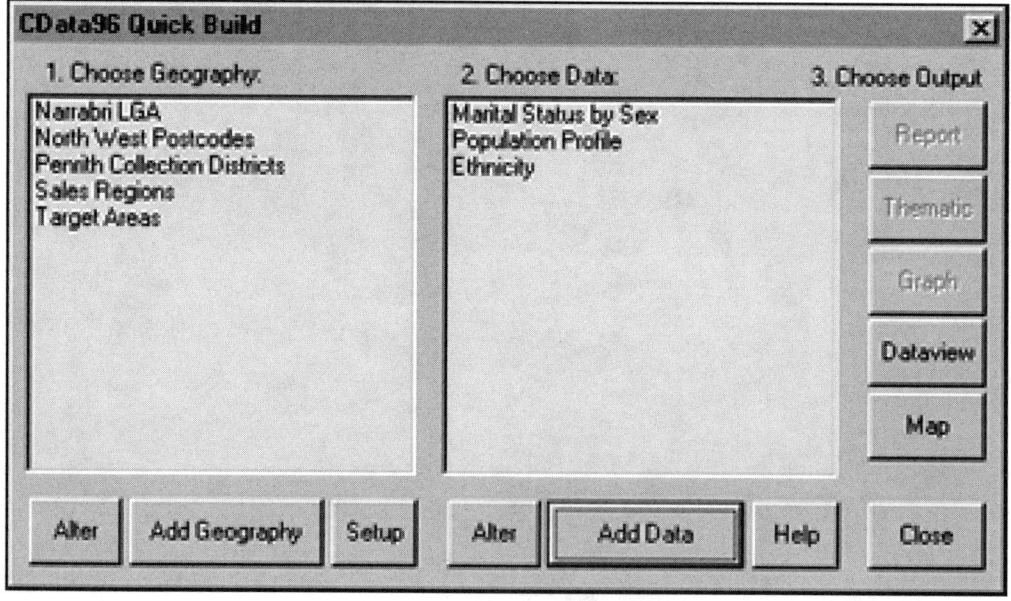

Quick Build window.

Rural Press can conduct analyses by searching for particular feature names such as schools, post offices, and hospitals. This functionality is particularly valuable when combined with ABS census data because the reports CDATA91 produces allow strategic marketing and sales planning that meet the needs of individual communities.

Innovations in CDATA96

The new ABS product CDATA96 builds on the strengths of its predecessor. CDATA96 offers a host of new features and 1996 census statistics. Consequently, it is expected to generate significant interest from current MapInfo users such as Rural Press.

Available on CD-ROM for Windows 95 and Windows NT, CDATA96 uses the powerful spatial analysis capabilities of MapInfo Professional's SQL query engine. The product is bundled with census collection districts and other statistical boundaries, and topographic data such as roads, major rivers, and railways.

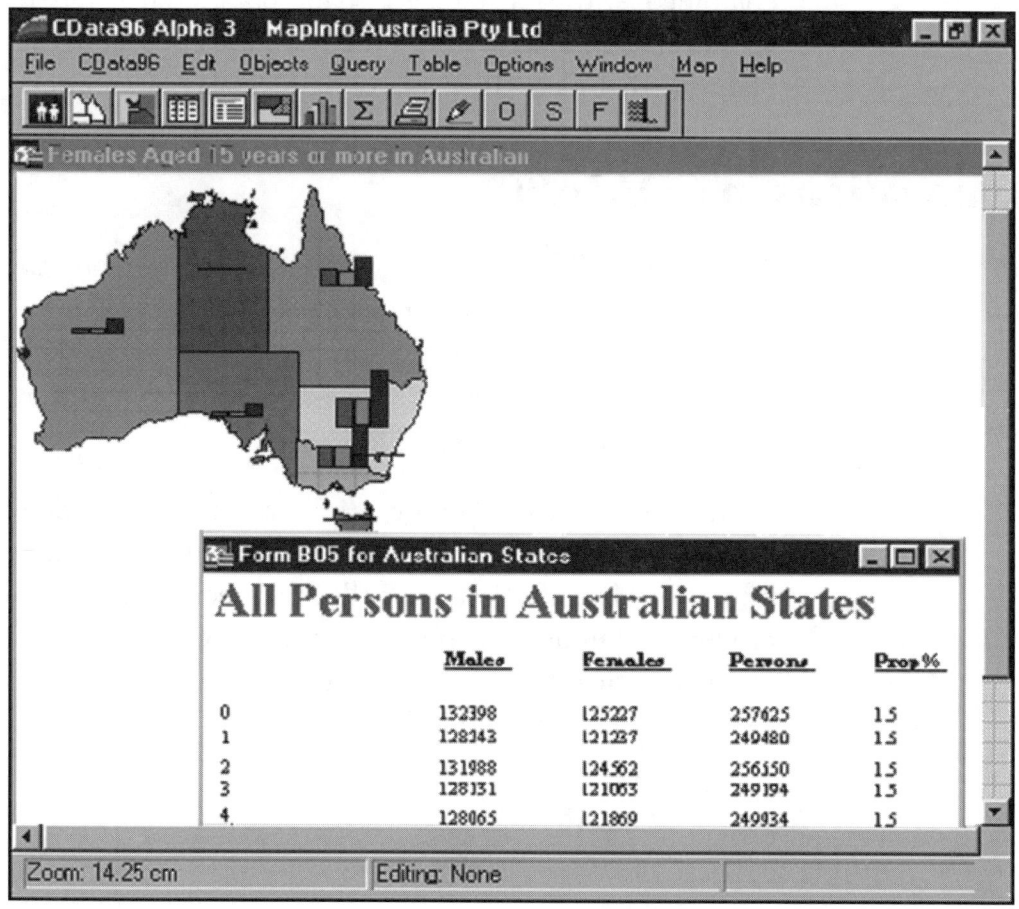

New data and functionality in CDATA96.

CDATA96 presents virtually every variable collected by ABS in the 1996 census in more than 50 tables, including data on age, sex, ethnicity, households, families, income, employment, religion, living arrangements, vehicles, and travel. All data are cross-tabulated to help users visually create a complete demographic analysis.

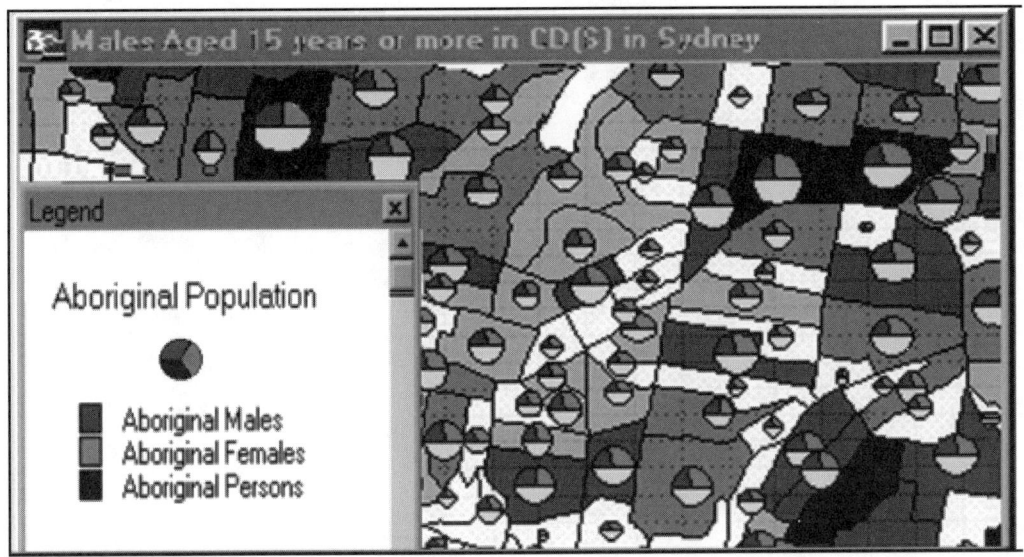

Pie thematic map of demographic data.

CDATA96 represents a powerful data management and analysis tool. Because census and corporate data can be analyzed together, users now have a vehicle for maintaining a competitive edge, discovering efficiencies, and creating other real business advantages.

Case Study
Retail Site Analysis • Vertical Mapper • Modeling

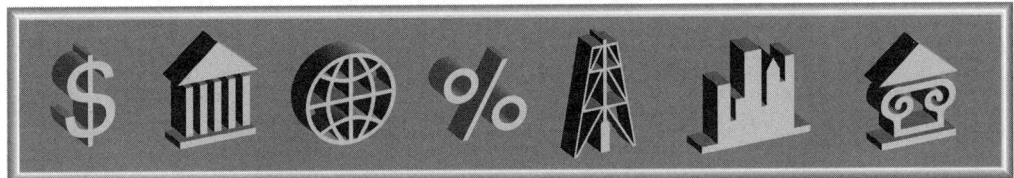

Canada Post

Since Confederation, Canada's postal service has been a cornerstone of the country's infrastructure, supporting economic activity and providing a vital communication link among all Canadians. Canadians entrust 10.9 billion letters and parcels to Canada Post Corporation every year—an average of 42 million pieces every business day—with the expectation that their business and personal mail needs will be served fairly and expeditiously.

To help meet this challenge, Canada Post has developed an extensive network of 7,500 retail stores. It is the responsibility of the corporation's Retail Analysis Department to ensure that each store is sited in the best location and that revenue predictions are accurate.

Methodology

To propose store locations and predict store revenues, Retail Analysis uses the following formal methodology.

1. Determine customer base
2. Establish store attractiveness factor
3. Define market area
4. Locate existing stores
5. Determine total market revenue potential
6. Create grid of store attractiveness
7. Establish optimal store locations
8. Perform gap analysis (existing versus optimal store locations) with field personnel
9. Generate patronage probability grid for the revised optimal store locations
10. Determine relative store strength by executing a point inspection of customer base against the patronage probability grid
11. Determine proposed store revenue using total market revenue potential and relative store strength

Each step is described in detail in the sections that follow.

Determine Customer Base

National exit surveys of Canada Post stores have established that customers can be divided into three categories.

Category	Definition	Revenue generation potential
Consumers	Customers who begin personal retail trips from home	40%
Businesses	Customers who begin business retail trips from business locations	47%
Employees	Customers who begin personal retail trips from business locations	13%

➥ *NOTE: Step 10 of the methodology, determining relative store strength, involves locating customer points of origin. Because business and "employee" customers have the same point of origin, the categories are grouped together, which means that separate geocoded databases of consumers and businesses must be developed for each store. A statistical review of Canada Post revenues against market variables has shown that revenues are more closely correlated to households than population. Consequently, the consumer database containing household—not population—data is compiled.*

Establish Store Attractiveness Factor

Store "attractiveness" is the measure by which a customer selects one store over another. This variable can be defined as total floor or shelf space, number of parking spaces, store age, or other combinations of elements defining its appeal. Because Canada Post's patronage is closely associated with a multi-destination retail trip, the density of retail activity surrounding a location is considered the best indicator of a store's attractiveness. This process is discussed further in the "Create Grid of Store Attractiveness" section that follows.

Define Market Area

Canada Post defines an urban market area as a single continuous urban area. The company has divided the country into subunits called Forward Sortation Areas (FSAs) for the purpose of organizing mail delivery. FSAs are identified by a unique, three-character code (e.g., K1A) that constitutes the first three elements in the postal code. Rural FSAs are distinguished from urban FSAs using a "0" (zero) for the second character. This process enables Canada Post to determine urban market areas using its own geographic definition.

Locate Existing Stores

Locating existing stores is obviously important when calculating patronage probability for new stores.

Determine Total Market Revenue Potential

When the market area has been defined and existing outlets located, a measure of the total revenue available in the market must be calculated.

Methodology

Create Grid of Store Attractiveness

CBI and SIC Data

The relative retail density of the market area is important when considering proposed locations for new stores. A compiled list of relevant retail stores and services, obtained from the Canadian Business Information (CBI) database, becomes the basis for calculating relative retail density. The term "relative retail" is defined using the standard industry code (SIC) applicable to each business and service found in the CBI. SIC categories representing businesses that consumers patronize regularly and frequently are given more weight than those patronized less frequently. The following map represents complete CBI information for a trade area together with a subset of retail business.

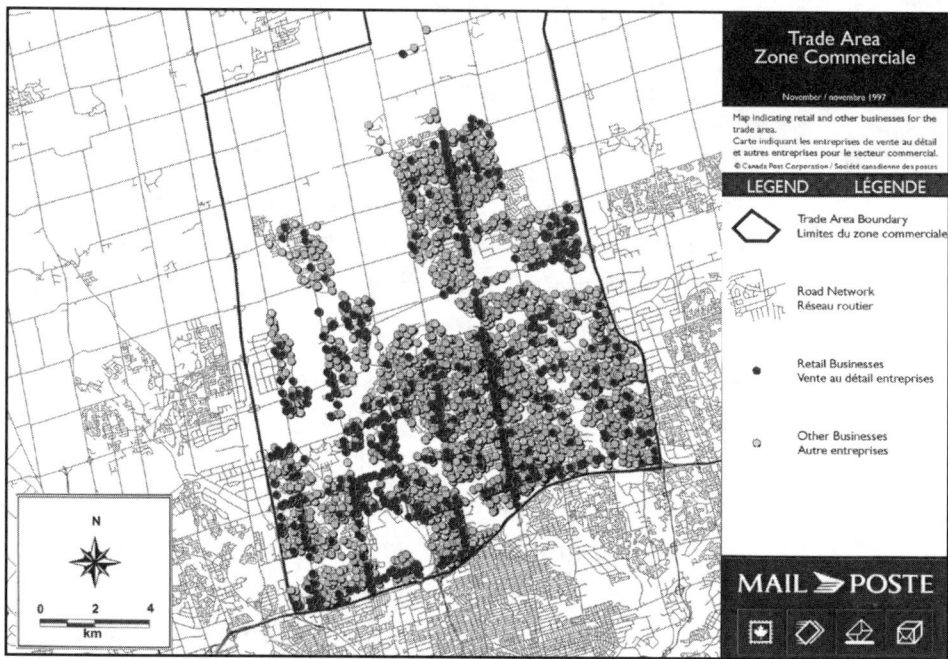

Trade area and retail business map.

Weighting Retail Areas

Retail SIC codes are weighted based on the frequency of return customer visits (frequent = 3, moderately frequent = 2, least frequent = 1). The rationale for evaluating retail sites in this manner is that the Canada Post customer is generally on a multi-destination retail trip. The more retail trips taking place in the area, the more attractive the Canada Post location. Retail point data are processed in Vertical Mapper using Simple Point Aggregation. A 250m coincident point distance is used to account for larger expanses that act as a retail district (e.g., intersections with street level retail businesses on both sides).

Methodology

This method also allows malls in heavy retail areas to be weighted more than secluded malls. Appearing below is a thematic map of the aggregated retail potential for the trade area depicted in the previous map.

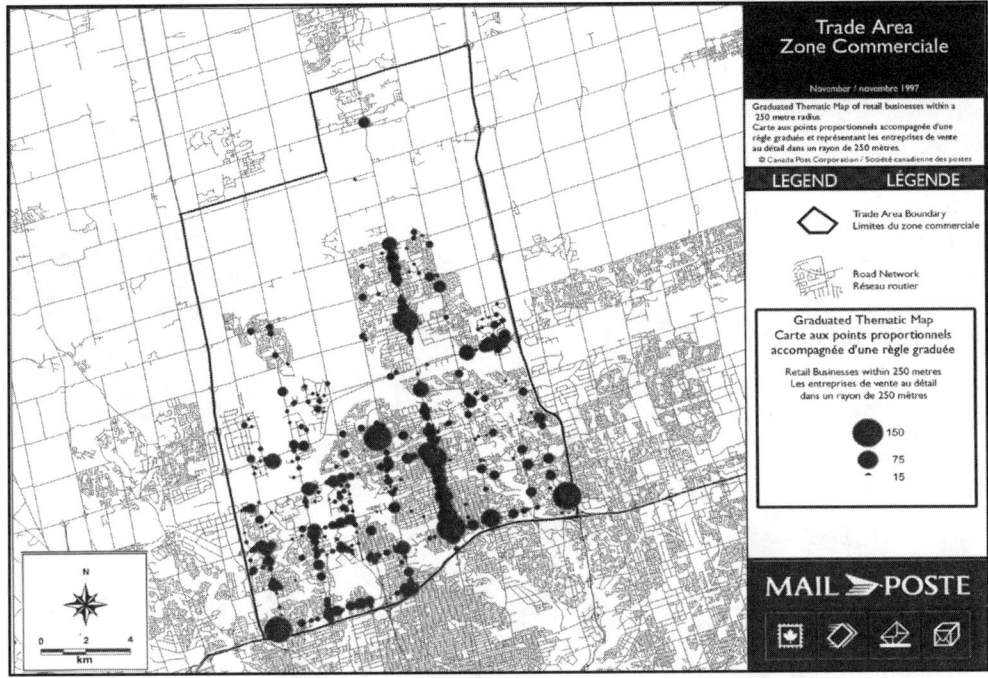

Thematic density map of retail potential for a given location.

Vertical Mapper

Vertical Mapper is then employed to create the retail density grid. The data are interpolated using inverse distance weighting (IDW). This method was selected over the others offered by Vertical Mapper (triangulation with smoothing, and rectangular interpolation) because the other options create a surface that passes through all points. In contrast, IDW does not "honor local high/low values" but does "estimate the local trends." Consequently, IDW distinguishes local retail trends and assigns relative values to each area. The grid surface that IDW creates serves as a guide to placing proposed locations in the most attractive area. High retail areas are represented in red (stars in the next image) with lower retail values progressing through a color gradient to blue, representing the lowest value (outlying dark gray tones).

Relative retail density calculated using Vertical Mapper.

Establish Optimal Store Locations

Areas with the highest retail density are identified as preferred site locations. Once the best locations are established, service level criteria such as drive time or maximum linear travel distance can be applied.

Perform Gap Analysis with Field Personnel

Every decision making process requires a validity check to reinforce that the assessment is correct. Performing a gap analysis with field personnel confirms accuracy of the data.

Generate Patronage Probability Grid

The Huff model, a gravity model supported by Vertical Mapper, creates a grid of patronage probability values indicating the likelihood that a customer will patronize a particular outlet. Two variations are available: the user can generate a probability grid for a single store (referred to here as the Single Huff model) or in a single pass compute patronage probability values for every store in the database and extract the maximum probability value at each grid location (the "Multi-Huff" model). In order to generate a Huff model, each store location must have a store attractiveness value field. When the attractiveness field is determined, the revised optimal store location file is processed through Vertical Mapper's point inspection with the store attractiveness grid. A visual representation of each outlet's actual patronage trade area is generated in Vertical Mapper using Multi-Huff. The resulting grid of maximum probability values effectively identifies areas where people are least adequately served, where great competition for customer business exists, and those retail sites that may be negatively affecting the revenue potential of adjacent locations. Here, the declining value is not retail value but patronage probability.

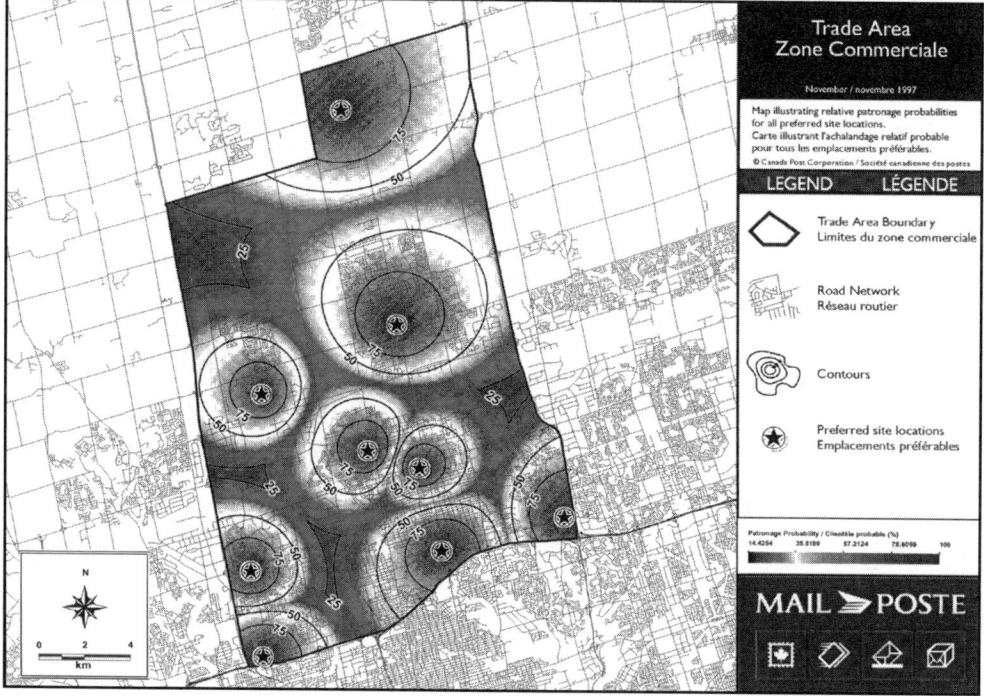

Multi-Huff model grid of patronage probability.

In order to generate proposed revenues, each optimal store location is modeled using the Single Huff model, resulting in patronage probability grids for each location. In this case, the locations of stores surrounding the target site influence the overall trend but patronage probability generally decreases as distance from the store increases.

Methodology

The grids allow analysts to inspect a particular store's patronage probability relative to the entire market.

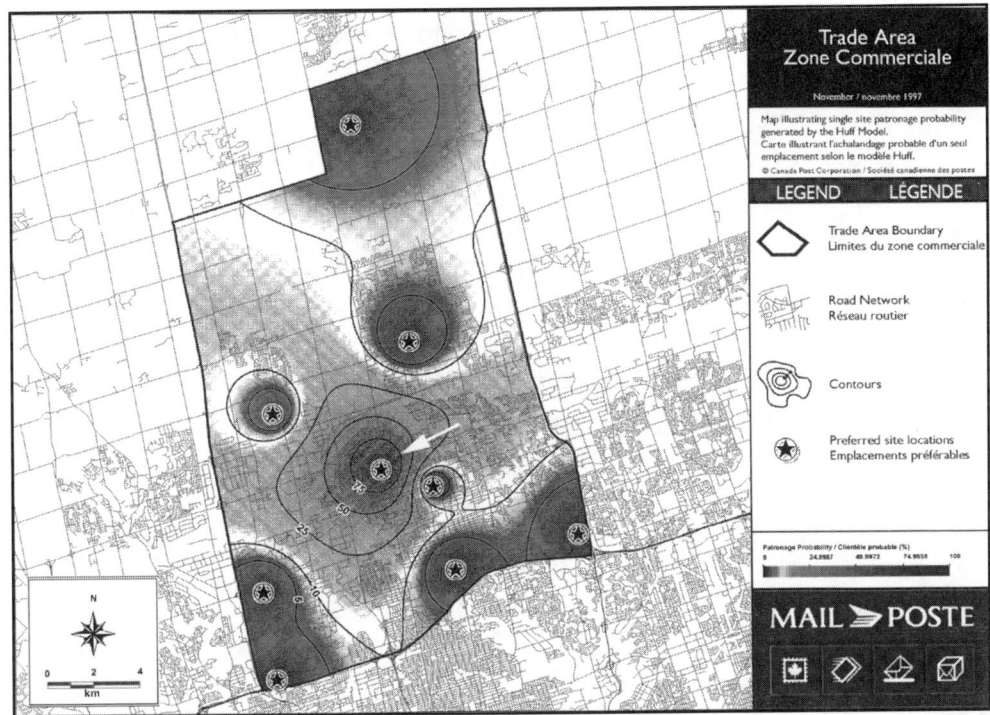

Single Huff model grid of patronage probability for a single store.

Determine Relative Store Strength

Relative store strength is determined by applying individual store patronage probability grids against the customer database using Vertical Mapper's point inspection. By multiplying the number of customers and the new patronage probability for each customer point, the store's strength can be expressed as a relative number of customers.

Determine Proposed Store Revenue

With every outlet in the market evaluated in terms of relative strength, revenues can be predicted for each outlet based on the total market dollar potential. For example, if a market contained three stores weighted 1, 3, and 6, respectively, calculating the total market dollar potential for each store would adhere to the following formula (assuming that total market potential equals $100):

$$1x + 3x + 6x = \$100$$
$$x = \$10$$

The potential revenues for the three stores are $10, $30, and $60, respectively.

> **NOTE:** *To learn more about Vertical Mapper, contact Northwood Geoscience Ltd. via the information listed in Appendix D.*

Conclusion

The store attractiveness grid established two elements: a visual representation of the best locations for a store and the relative strength of all store locations (both actual and proposed). Store location and relative strength are the variables required to use the Huff model. The Huff model does not predict revenues but produces a patronage probability grid. The patronage probability grid must be applied against the customer database to establish the total number of store customers. With a fixed dollar value per customer, potential store revenue can be estimated. In this manner, sites for new stores and their potential to generate revenue can be evaluated.

Case Study
Routing • Information Systems

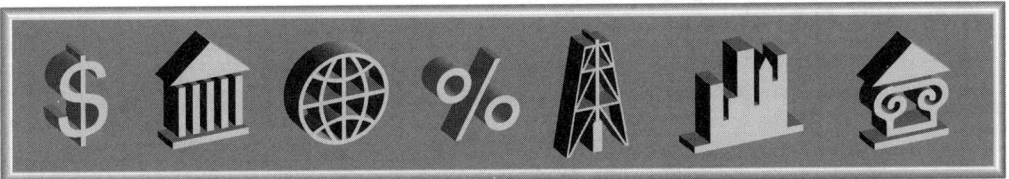

Emergency Medical Service Institute

In southwestern Pennsylvania, more than 500,000 emergency and non-emergency ambulance calls occur each year. Ambulance services licensed by the commonwealth's Department of Health are under regulatory guidelines to respond to emergency calls within 10 minutes in urban abd suburban areas. (In rural areas, the guideline is 20 minutes, if they are providing advanced life support backup to a basic life support ambulance.)

The Emergency Medical Service Institute (EMSI) is the largest of 16 regional planning agencies in the commonwealth that evaluate the ability of ambulance services to fulfill the guidelines. A non-profit agency located in Pittsburgh, EMSI oversees more than 200 ambulance services in a 10-county area spanning 7,000 sq mi and serving 2.7 million people.

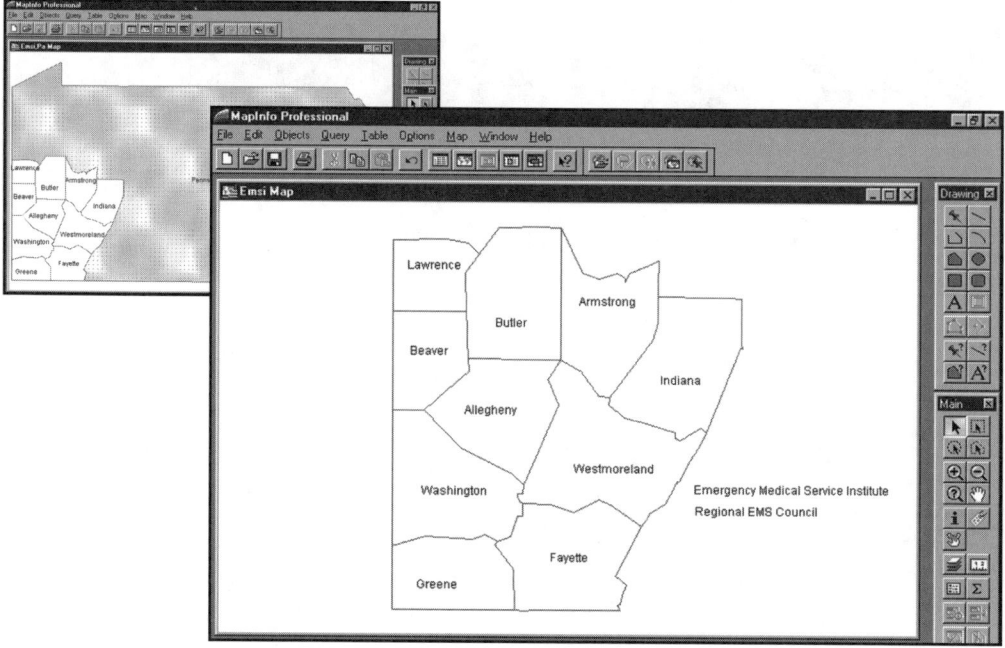

EMSI's service territory.

EMSI's mission is to plan, develop, maintain, expand, and improve emergency medical service systems within its territory. Most activities are regulatory in nature, such as EMS training institute and medical command facility accreditation, administering rescue and EMS training program certification exams, and coordinating communications among service providers and counties in the region.

Desktop mapping solutions from MapInfo Corporation and On Target Mapping are extensively used to support response time evaluation and conduct license inspections.

Response Time

In the past, response time evaluations were extremely time-consuming for EMSI staff, who physically drove sample routes for each ambulance service to clock and test response coverage. Following two years of in-field testing by EMSI, On Target Mapping (also located in Pittsburgh) developed a software solution geared for fire, police, hazardous material, ambulance, and paramedic departments. Incorporated into On Target's DRIVE PLUS version 5.0 software, the functionality allows EMSI to perform a complete analysis of hypothetical response scenarios for an emergency. Using real street distances and speeds, the program calculates response times from numerous points, maximum coverage areas, and appropriate stations to respond based on travel time and location.

DRIVE was developed to respond to real world variables (e.g., road condition and type, weather, time of day, and vehicle type). Up to 36 variables allow users to create a customized, realistic representation of the site commute. In addition, MapEdit can be used to determine road closures and one-way restrictions. (DRIVE software is used for single site analysis, and DRIVE PLUS for multiple sites.)

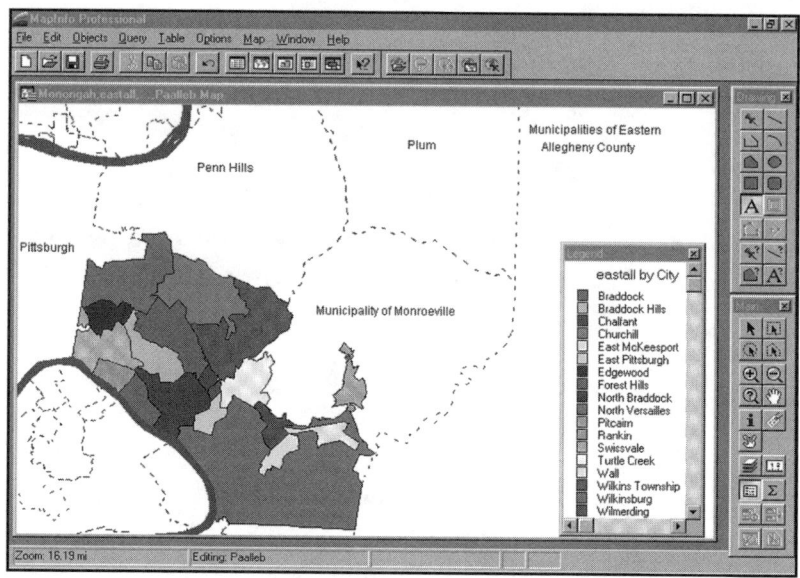

Communities southeast of Pittsburgh whose emergency response services are overseen by EMSI.

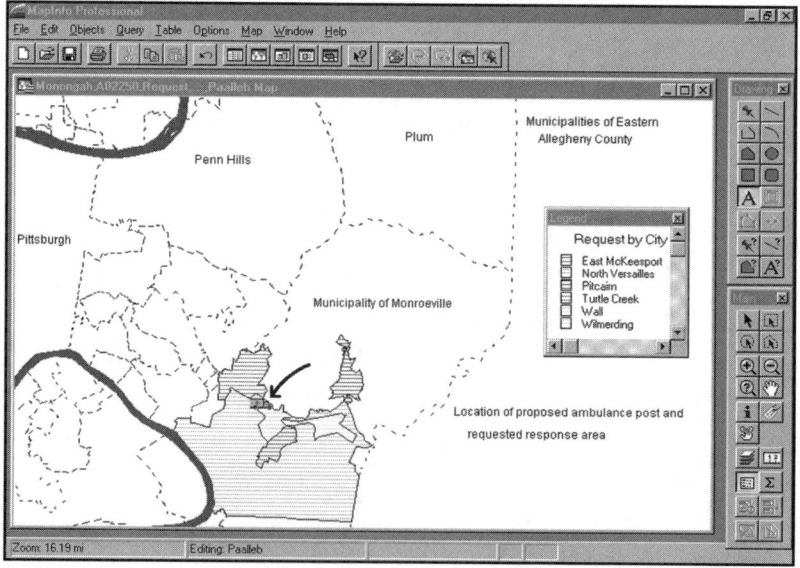

Location of a proposed ambulance post and the requested response area. The black, curved lines on the upper and lower left side of the image show the Allegheny and Monongahela Rivers, respectively.

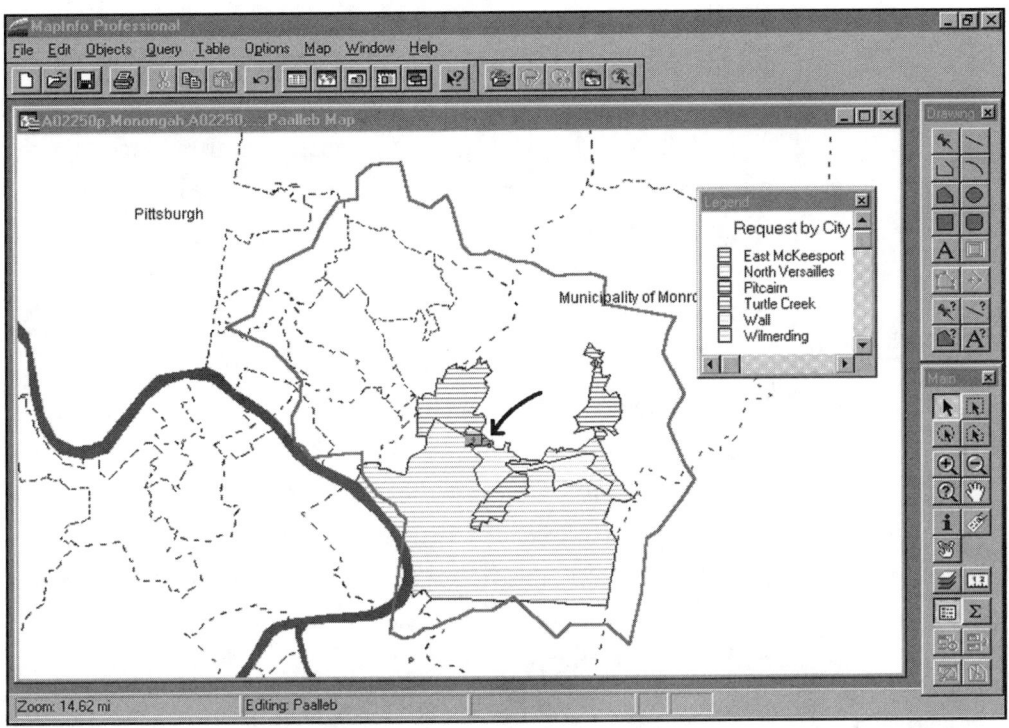

Ten-minute response polygon for the proposed ambulance post. All requested communities would be assigned to this post based on the DRIVE analysis.

The mapping solution is particularly well suited for evaluating response times in the region's densely populated urban areas. In addition, it makes quick work of road network restrictions created by mountainous terrain, including tunnels, rivers, and bridges in Allegheny County (which contains more bridges than any other county in the nation). Finally, the software helps EMSI compensate for the lack of a single 911 system. For example, because Allegheny County—also the most populous county in the region—has not yet instituted a countywide 911 system, the responsibility for receiving and dispatching ambulance services is spread across numerous municipalities.

EMSI also uses the software to provide technical assistance to rescue-service providers within the region. It can assist them in determining the number of vehicles required to adequately service constituents within their area, analyze mergers, and make recommendations regarding additional ambulance-service providers.

Licensing

In the area of license inspections, mapping solutions enable EMSI staff to shorten evaluation time of response areas and increase productivity. Most evaluations are conducted in 20 to 30 minutes, instead of several hours. EMSI can take traffic, weather, and vehicle types into account, and then print out detailed maps that graphically display the results.

Conclusion

In addition to response time and licensing evaluations, EMSI uses the software to identify underserved areas, track locations of medical facilities such as hospitals and medical command facilities, and create a database of training program locations. EMSI is also in the early stages of cataloging fire departments that provide rescue services. Such user friendly software saves EMSI significant time and effort as it fulfills its regulatory mission.

Acknowledgment

OnWord Press thanks *Earth Observation Magazine*, which published an earlier version of this case study, for granting reprint permission.

CASE STUDY
TELECOM • MARKETING • INTERNET

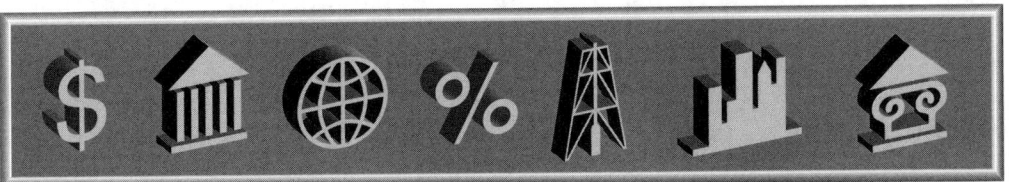

PageNet

Public Safety Associates, Inc. (PSA), was founded in 1988 in Dallas, Texas. With extensive experience in GIS and public safety, PSA specializes in telecommunications, public safety, and petroleum solutions with desktop mapping and public safety applications.

Also based in Dallas, PageNet is the world's largest wireless messaging company, serving more than 10 million customers with a network that reaches 90% of the U.S. population. The company also has subscribers in Canada, Brazil, Spain, the U.S. Virgin Islands, and Puerto Rico.

With more than 9,500 transmitters across the country, PageNet owns and operates the nation's most extensive wireless communications network. It provides wireless data communications for consumers and a variety of business segments, including mobile computing, telemetry, field sales and services, public safety, and transportation. In addition to numeric and alphanumeric paging, PageNet provides news and stock updates, voicemail, and fax forwarding and wireless data transmission to palmtop and laptop computers and personal digital assistants.

Defining the Problem

PageNet defined three problems for PSA to solve using the GIS capabilities of MapInfo: create a central repository of up-to-date information regarding PageNet's architecture for the company's executives; develop a tool for quickly and easily analyzing potential new tower sites for marketing staff; and create a centralized database of tower site information for engineers.

The first problem—lack of current data for executive decision making—was a direct result of the rapid growth of the telecommunications industry. Demand for service had grown so quickly that printed site and marketing data became obsolete. The second and third challenges, evaluating sites for new towers and cataloging data on current ones, were well suited to GIS because tower data and coverage areas are geographic in nature.

Solving the Problem

To meet the needs of PageNet, PSA developed PCS SiteManager, an interactive system containing personal communication system (PCS) marketing and radio frequency (RF) engineering data. Designed for use via a corporate intranet or the Internet, PCS SiteManager was developed using MapInfo ProServer (since replaced by MapXtreme), MapX, and MapXsite technology. (For further explanation of each product, see Appendix B, "MapInfo Product Line Overview.") While a lengthy case study could be devoted to the programming aspects of this project alone, four programming languages were used to incorporate the three distinct modules listed below into PCS SiteManager.

- Marketing/Executive Level Overview Module
- Site Reporter Module
- Real Time Site Analysis Probe

PCS SiteManager was designed to keep executives informed and provide a master set of coverage maps and demographic and site data in a single location, in order to reduce duplication and prevent field technicians from using old information. It was also designed to save time by automating market analysis of existing and proposed systems.

Launching PCS SiteManager

Executives, marketers, and engineers with access to PCS SiteManager access the following screen when launching the software.

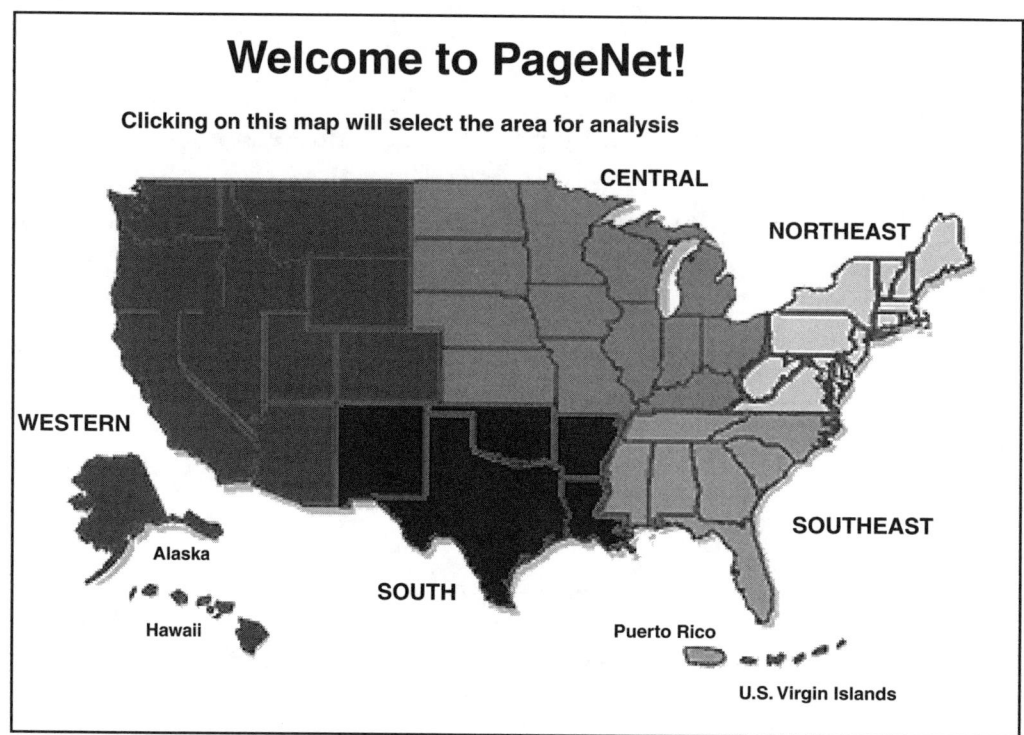

Opening Web page.

Clicking on a region generates a page tailored to the selected area. For each region, the user can perform the following operations.

- View capacity reports
- Perform a demographic analysis
- View current RF coverage predictions
- Create a hypothetical network and show its availability
- View current performance statistics

Marketing/Executive Level Overview Module

The Marketing/Executive Level Overview Module compares existing area coverage with area demographics on a census tract level. Fields available include total population, total household units, median income, and other desired customer demographics. The module can also report demographics not currently covered within the system's configuration. This feature is especially useful for evaluating diminishing returns on additional cell sites.

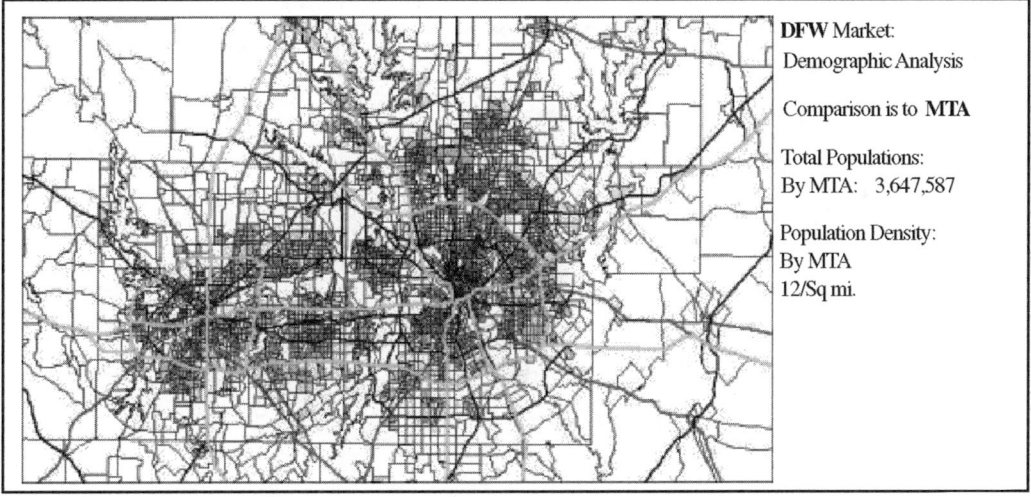

Compare coverage to demographics.

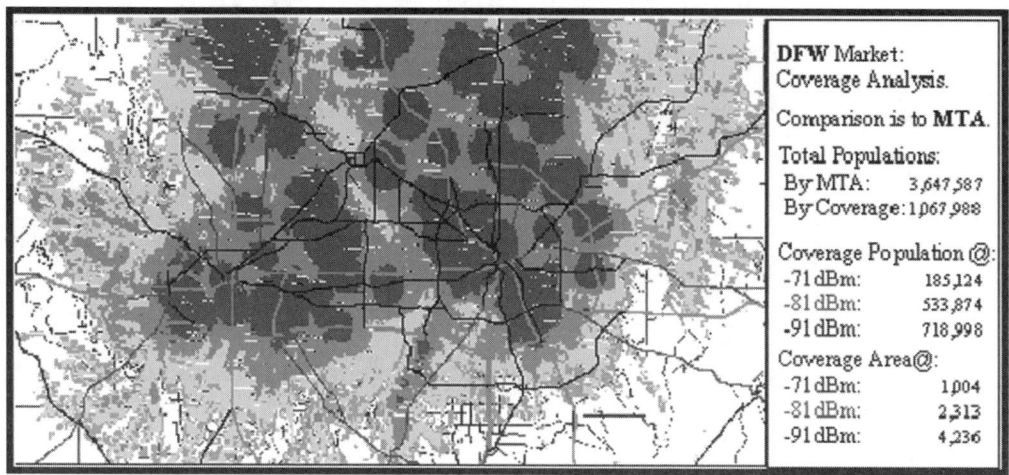

View capacity reports.

Site Reporter Module

The Site Reporter Module provides current real time information on individual PCS sites. Customer defined demographics (e.g., population, income, age, household, and so forth) and average daily drive times—the number of potential customers who drive through a given area on an average day—are available.

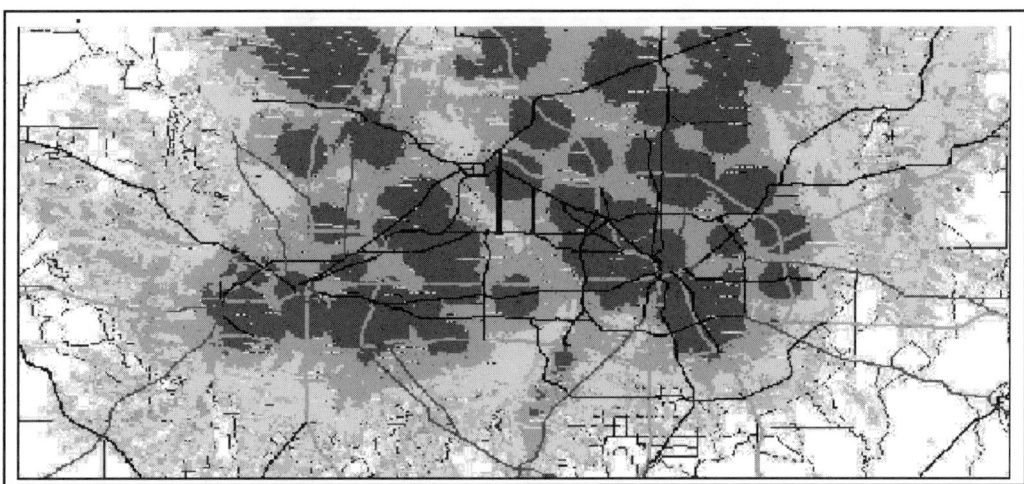

Show RF coverage for a site.

When a site is chosen, a table listing the site and coverage information is displayed below the image.

Real Time Site Analysis Probe

The Real Time Site Analysis Probe interrogates each PCS site in order to determine zone, transmitter, receiver, channel, and terminal status. The probe interacts with customer provided site information in a real time or archived environment.

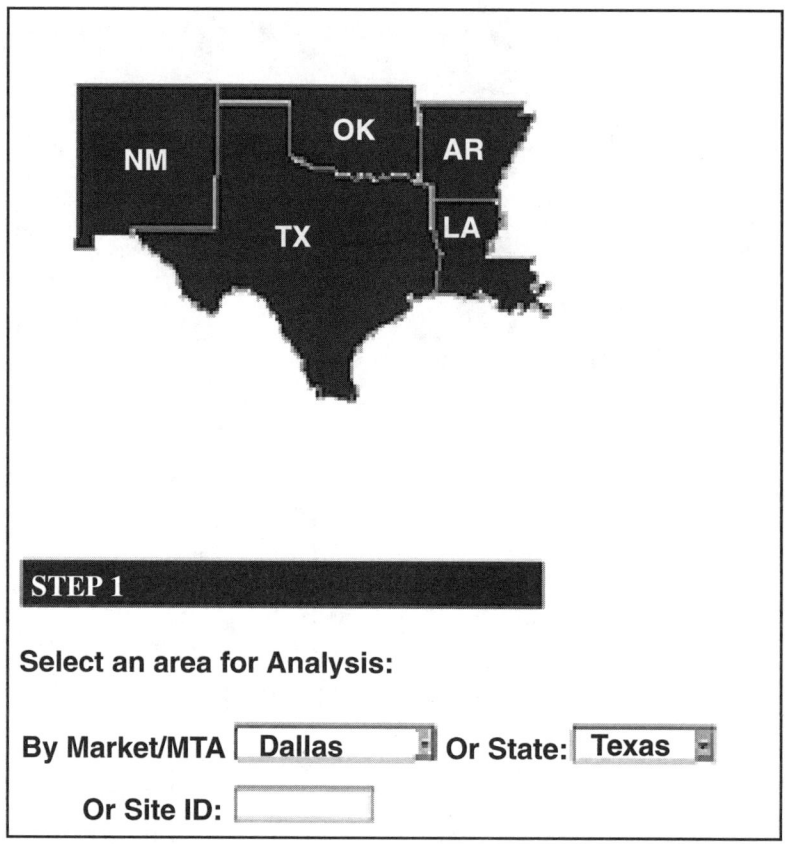

Regional start page in the Probe.

The user selects an area for analysis (or types in a site identification number), chooses a specific report (see the next image), and waits for the system to generate the report. The module allows the user to view demographic, RF coverage, or virtually any real time statistics. Based on the amount of information provided by the user, the system will select the appropriate level of detail. For example, choosing Texas yields less detail than selecting Austin or site CNT-523.

Solving the Problem

```
┌─────────────────────────────────────────────────────────────┐
│ ■ STEP 2                                                    │
│                                                             │
│ Select Analysis to Display:                                 │
│                                                             │
│ ⦿ Demographic Area Analysis, using:  [Daytime Population ▾] │
│                                                             │
│ ○ RF Coverage Area                                          │
│                                                             │
│ ○ RF/Capacity Plan                                          │
│                                                             │
│ ○ Statistical and Performance Reports, showing  [Transmitter Status ▾] │
│                                                             │
│ ○ Custom Demographic/RF Coverage Area Analysis              │
│                                                             │
│ ○ Custom Site Builder/Capacity Analyser                     │
│                                                             │
│ If you have selected the wrong options, simply click the Reset button below. │
│                                                             │
│ ■ STEP 3: GO!                                               │
│        [ Search Database ]  [ Reset ]                       │
└─────────────────────────────────────────────────────────────┘
```

Steps 2 and 3 in the Probe.

Options in Step 2 of the Probe are described in more detail below.

- **Demographic Area Analysis.** Demographic data, such as ethnicity, gender, and income, can be broken down into daytime and residential categories.
- **RF Coverage Area.** Shows coverage for the site.

- **RF/Capacity Plan**. Information regarding how the site was designed to work.

- **Statistical and Performance Reports**. Displays near real time operational statistics for the site.

- **Custom Demographic/RF Coverage Area Analysis**. Compares actual coverage with demographics to uncover potential problems or missed customers.

- **Custom Site Builder/Capacity Analyser**. Estimates a virtual system.

Future Enhancements

As with most projects, when PageNet began using PCS SiteManager the company wanted greater functionality. Forthcoming additions to PCS SiteManager are described below.

Obstacle Module

The Obstacle Module, which will show building locations and heights for proposed coverage areas, will be used to evaluate current system design and new construction.

Site Comparator Module

The Site Comparator Module will contrast demographics, propagation (the physical distance radio waves travel), and drive times for two existing transmitter sites. The module will also return new values for hypothetical sites to assess whether a new site would serve customers more effectively.

Future Enhancements

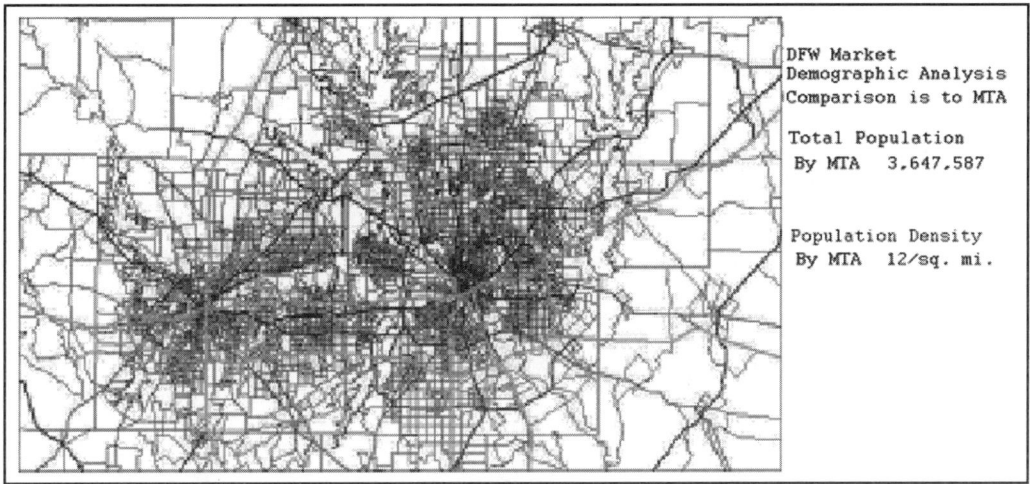

Site Comparator Module screen for Dallas-Ft. Worth.

When an area is selected, the right side of the screen provides a summary of the selected area. Textual information can include residential versus daytime population, ethnicity, age, and income. Future iterations of the product will generate custom reports and display points of interest.

CASE STUDY
INFORMATION SYSTEMS

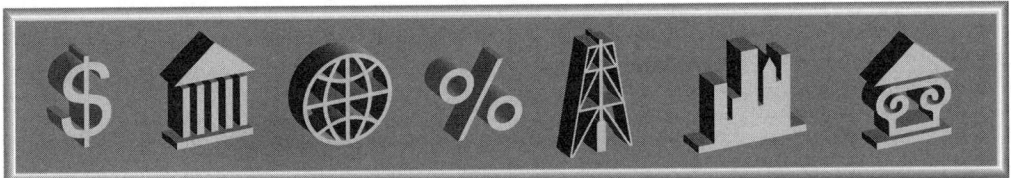

LOUIS DREYFUS NATURAL GAS

Louis Dreyfus Natural Gas Company is a rapidly growing domestic energy company focused on the acquisition, development, exploration, production, and marketing of natural gas and crude oil. Louis Dreyfus Natural Gas has an interest in 1.4 million gross acres of leasehold and 6,500 producing wells with significant exploration and exploitation potential in the Mid-Continent, Permian, Gulf Coast, and Sonora regions in Texas and New Mexico. High quality oil and gas basins with numerous geological formations likely to contain hydrocarbons are concentrated in these regions.

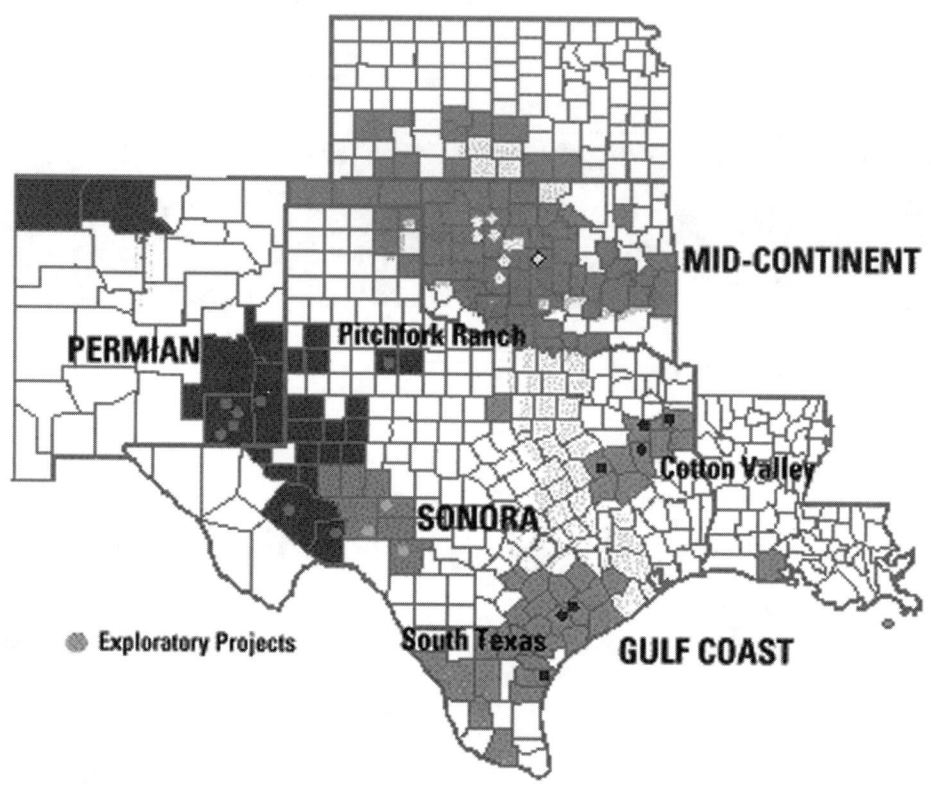

Louis Dreyfus lease locations.

Louis Dreyfus' proven reserves have grown at a rate of 36 percent per annum over the past five years, reaching the current production rate of 1 trillion cu ft equivalent.

The company's drilling strategy for development and exploratory efforts is straightforward. Engineers assimilate and evaluate as much data as possible, and drill in areas likely to contain multiple oil and gas bearing formations in order to minimize dry hole risk. Louis Dreyfus looks for opportunities in traditional development areas by drilling deeper for new horizons at minimal incremental costs, often finding overlooked opportunities in mature producing basins.

The company's teams of geologists, engineers, and land professionals collaborate on projects using various technologies, including 3D seismic and computer-aided exploration mapping, to gather, analyze, and interpret data and maximize exploration opportunities.

Problem Definition

The company employs a host of applications developed to meet the needs of the oil and gas industry. In particular, one application that plots well sites to contoured sub-earth geological layers lacks sufficient mapping capabilities.

Several years ago Louis Dreyfus purchased MapInfo Professional for the first time. Users were pleased with the software's ease of use and high quality maps. The company uses the software to geographically display lease information and oil and gas well locations. The maps depict the location of each well according to section, township, and range coordinates and in relation to other wells. The software was such a success that walls throughout the company were papered with new maps in a matter of months after the initial purchase.

One particularly significant map shows reservoirs for oil and/or gas production wells in specific areas. This map is extremely important for prospecting engineers and geologists evaluating new drilling locations. As part of this application the company created a map known as a 9-section plat, which is used in various situations from requesting drilling permits to legal issues.

9-section Plat Maps

A primary goal of petroleum engineers and geologists is to exploit existing oil and gas producing properties already owned by the company. To achieve this goal, existing well information for a given property (or adjacent to it) is studied to determine if more oil and gas can be extracted. One type of map used in this process, a production map, is used to show where wells are located and the reservoirs from which they produce. The following map shows well locations and a legend indicating the source reservoir. When the map is reproduced in color on a computer monitor or printer, each reservoir is quite distinguishable.

9-section Plat Maps

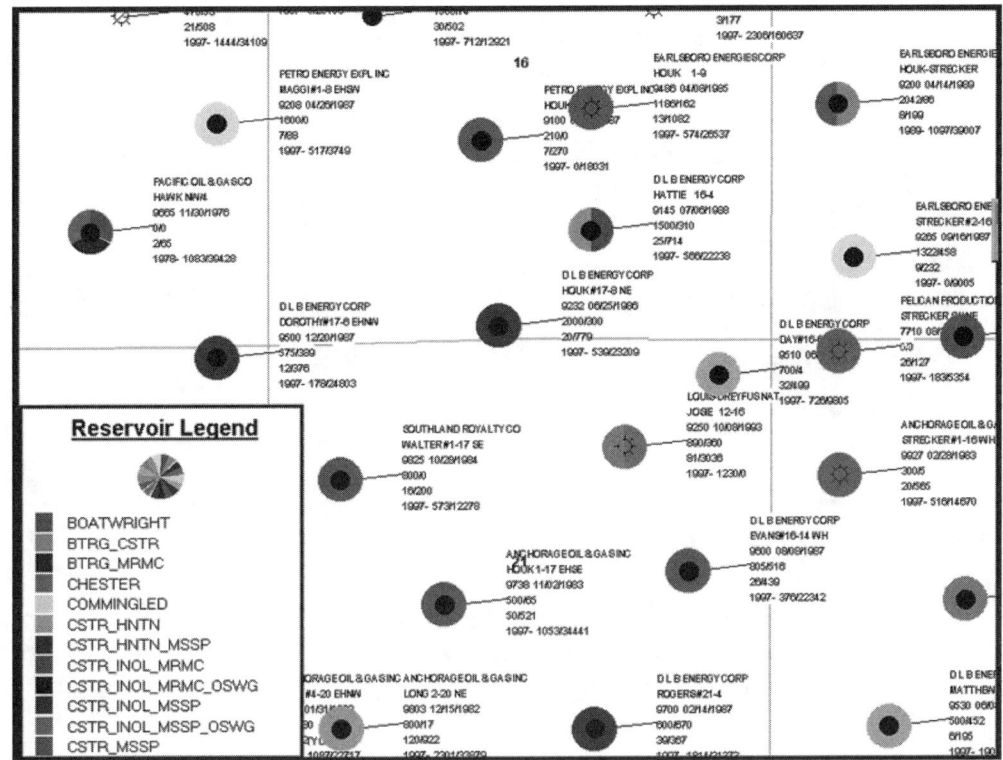

Production map and reservoir legend.

The map takes advantage of MapInfo's functionality for creating pie thematic maps to identify reservoirs. Many wells in this example have only one color, indicating that they are producing from a single reservoir. The next illustration zooms in on a well producing from more than one reservoir. Other information regarding the well is also displayed.

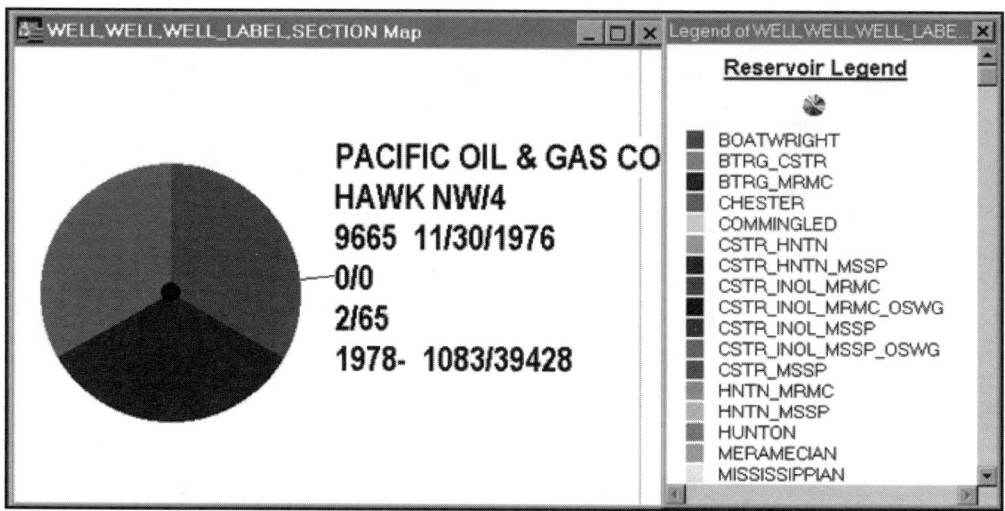

Detail of a well that draws from more than one reservoir.

The Pacific Oil & Gas Company well in the above illustration is producing from three reservoirs: Boatwright, BTRG_MRMC, and Chester.

Data Files

Creating 9-section plats requires several pieces of data: a graphic data table of relevant section/township/range regions, and oil and gas well information (including latitude and longitude coordinates and source reservoirs).

Section/Township/Range Data

Records concerning properties are legally described using unique section, township, and range identifiers. For example, a piece of farm property is described as the southeast corner of section 4, township 28N, range 22E. Oil and gas wells are located with the same type of legal description or according to latitude and longitude.

9-section Plat Maps

Vendors of section/township/range data include Tobin Data Graphics, Petroleum Information/Dwights, Topographic Mapping, and Whitestar. Data from Tobin are organized into two tables of township/range and section shapes. In the following illustration bold lines represent township/range shapes and thinner lines depict sections, which are labeled according to respective unique numbers.

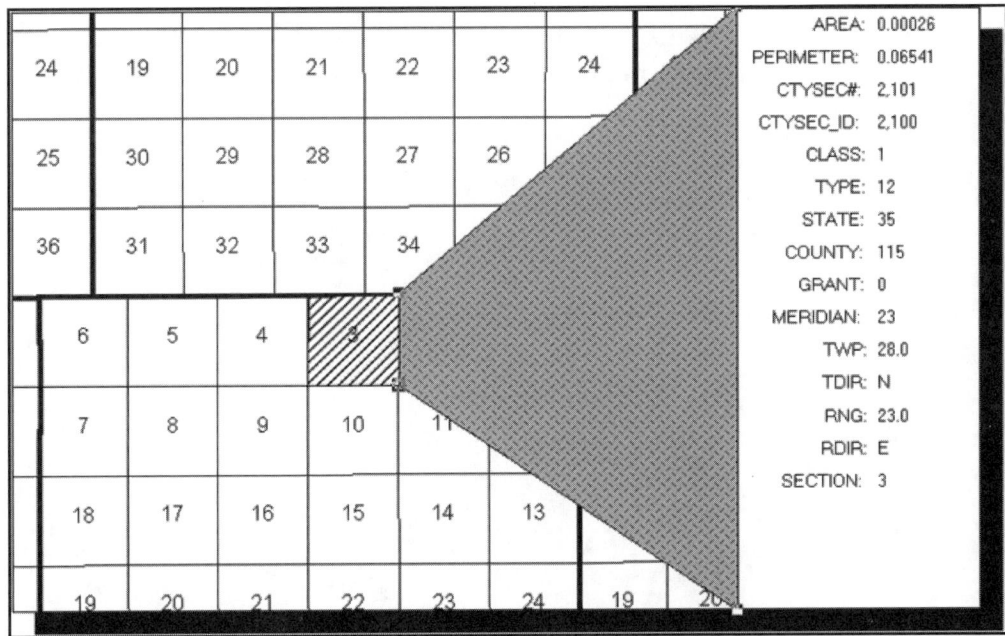

Section/township/range data.

Oil and Gas Well Data

Because oil and gas well data must be filed with a state's corporation commission (and are therefore public record) they can be obtained directly from a government agency. However, some private vendors have also begun marketing value added gas and well data. One company, Petroleum Information (PI)/Dwights, has built a reputation for accurate data and as such may be a worthwhile alternative to public information, the accuracy of which can vary.

PI/Dwights maintains a massive well and production database of more than 2.6 million U.S. wells. The sample data shown in the following images were acquired in DBF (or dBase file) format. The data come in two files, a scout ticket file (containing well name, location, and ownership) and a production data file (containing well production and reservoir statistics).

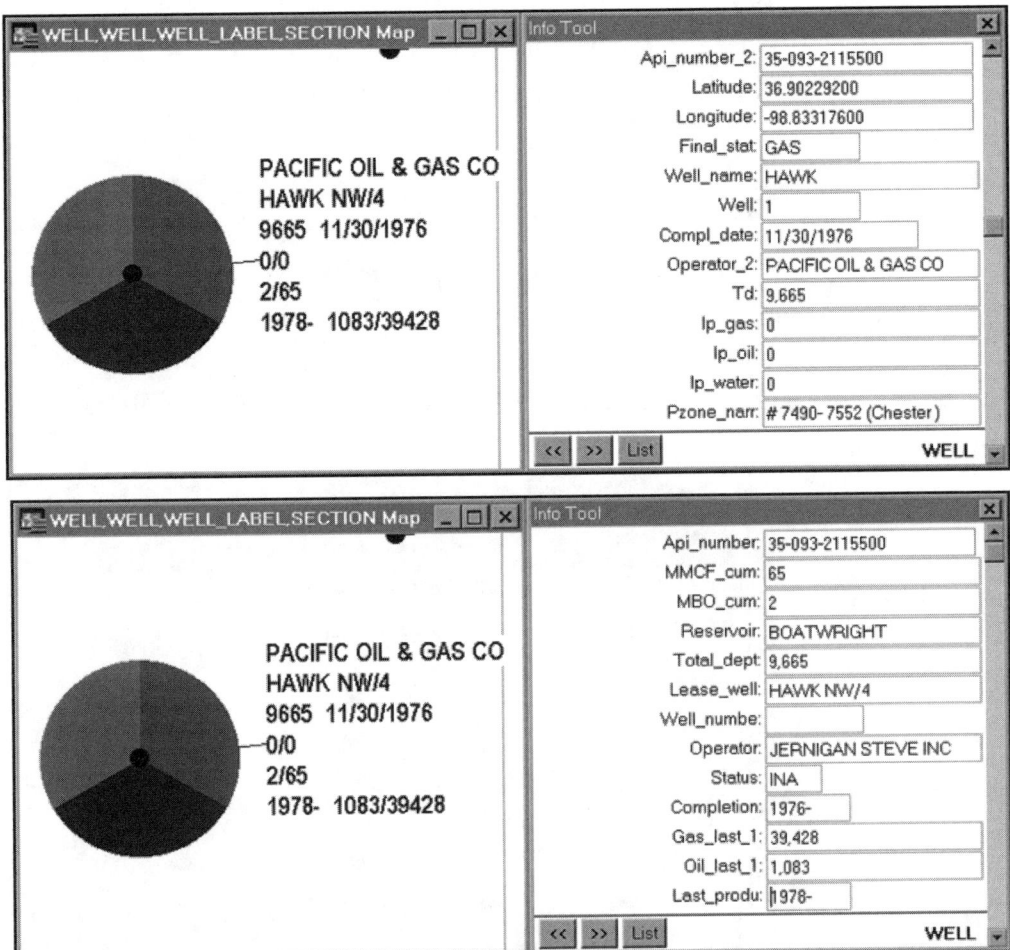

Scout ticket (top) and production data (bottom) files.

Creating the Map

Louis Dreyfus contracted IntelleVue to write a MapBasic program that would ease the creation of oil and gas maps in MapInfo. The program ultimately automated several hours of manual steps previously required to produce adequate maps. Because many of the processes were repetitive, program automation was worthwhile.

Oil and Gas Well Data

The first step in the program is preparing oil and gas data tables for the map. Data received from PI/Dwights are not in a format suitable for mapping. Once the user chooses the data tables to process, a dialog box prompts the user to select reservoirs used in the pie thematic.

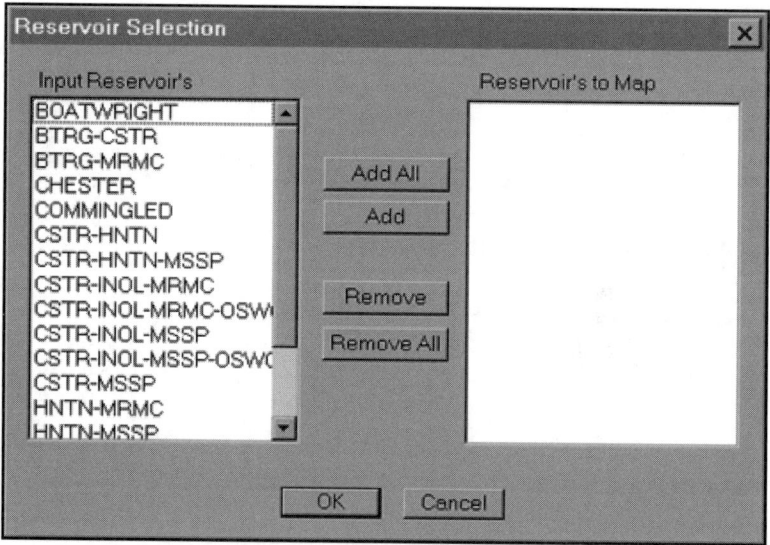

Reservoir Selection dialog box.

Once reservoirs are selected, the program automatically builds the necessary data tables. Next, the program creates the Map window, pie thematic, and Legend window. Each well point in the Map window is appropriately labeled.

Creating the Map 223

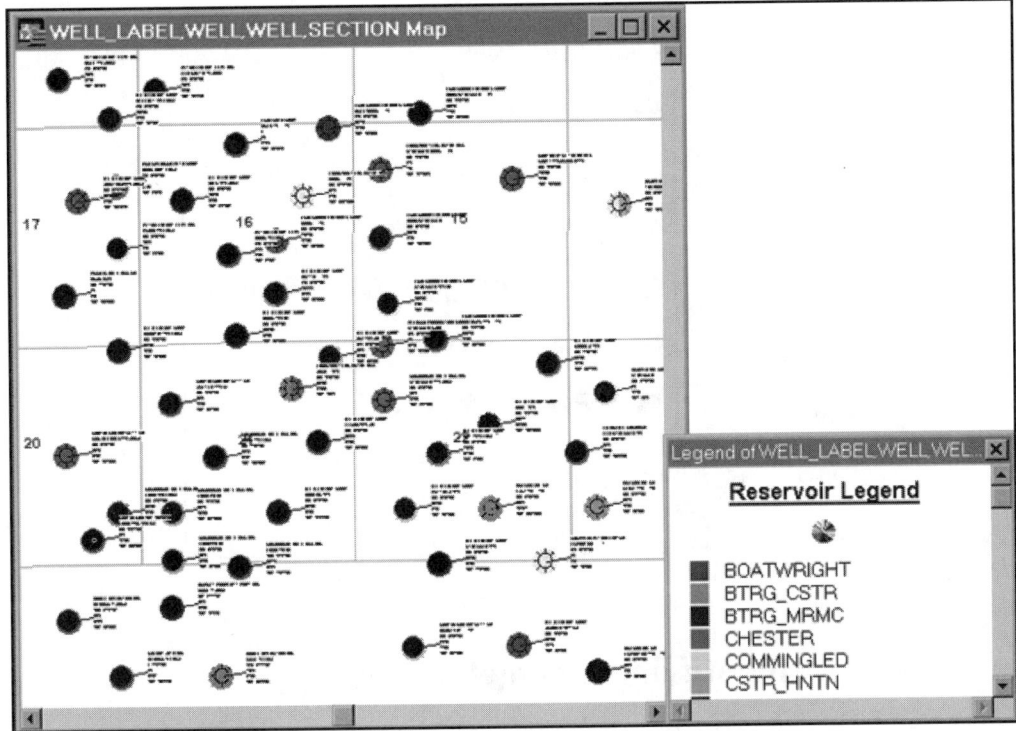

Map window, pie thematic, and Legend window.

A status column in the well data file directs the software to label each well point with the correct MapInfo oil and gas symbol (see the next image).

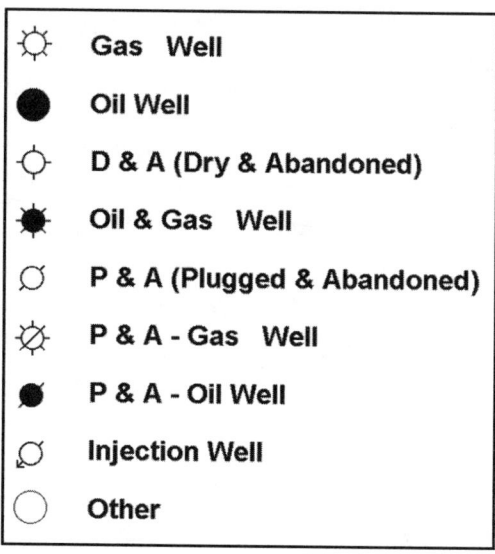

Oil and gas well legend.

Identifying a Target Section

At this point the user may need to adjust label positioning on the map. Once this step is complete, the program creates the Layout window for the map to prepare it for printing. The user issues a command to generate a 9-section plat and is prompted for a target section. Once the user selects township and range values, valid section numbers are displayed. The next dialog box makes it easy for a user to correctly select the target section.

Creating the Map

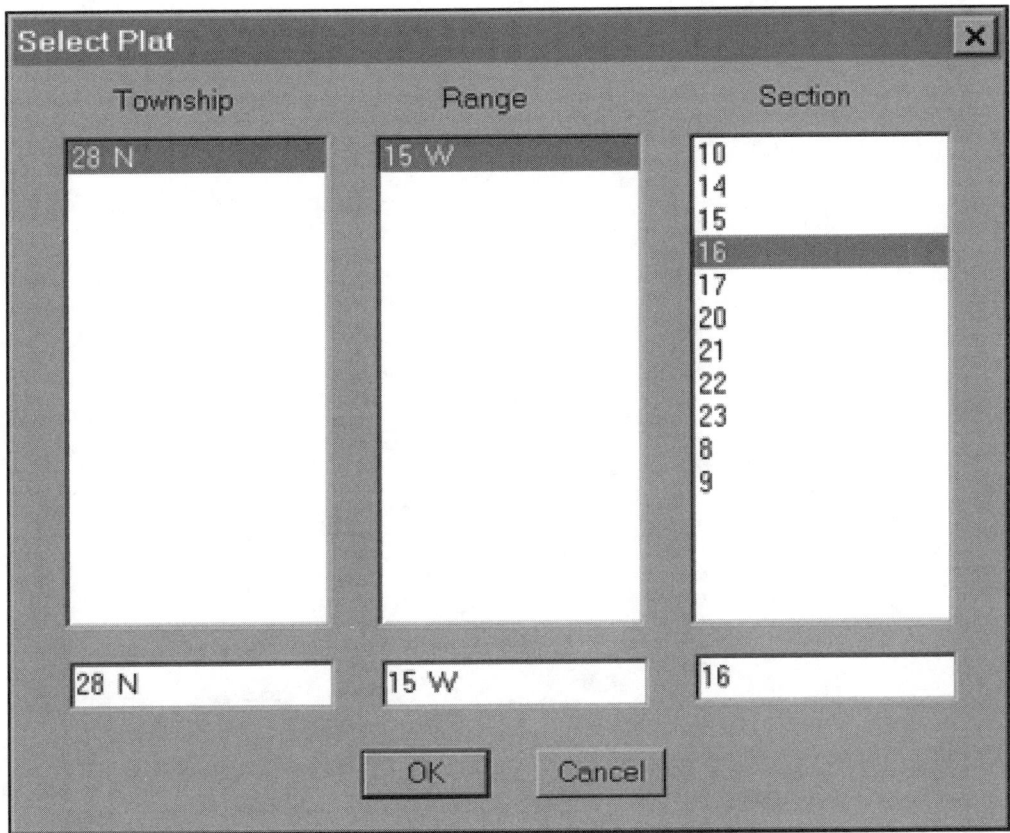

Select Plat dialog box.

After the user clicks on OK, the program automatically builds the layout for the 9-section plat map. A company logo and title block are added, as well as township and range labels.

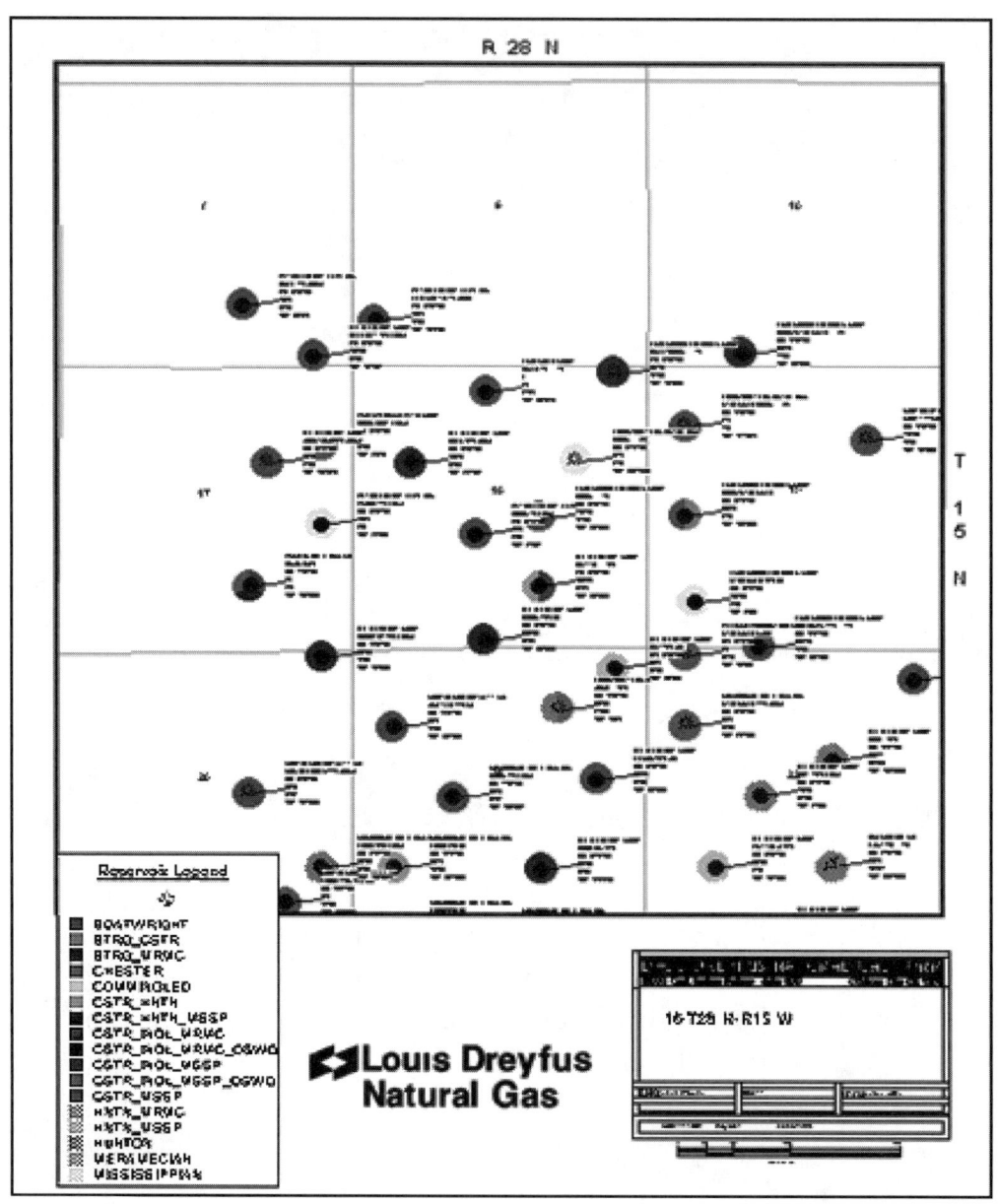

Louis Dreyfus 9-section plat map.

At this point, the user is again allowed to adjust titles or other positioning. Compared with manually creating the plat, such editing is minor. In fact, when a company manager was shown the automation available through MapInfo, he requested that it be available for all engineers—a significant time saver for many individuals in the organization.

Other Applications

Louis Dreyfus also uses MapInfo (in conjunction with Vertical Mapper) to generate contours depicting reservoir characteristics, well production rates, ultimate recoverable oil and gas reserves, and other data. The company was especially pleased to discover that MapInfo could handle very large maps with large amounts of data, unlike other software packages. For example, MapInfo is the only software capable of accommodating boundary and well data for the entire state of Oklahoma using 32-bit processing. Consequently, far more requests for very large maps are surfacing. In addition, MapInfo is used to display the company's subscription coverage of relevant production and well data purchased from PI/Dwights. Staff can build maps to delineate areas of responsibility. Finally, when the company is considering an acquisition, MapInfo shows well locations and property ownership data for both parties.

The list of applications is long and continues to grow. At a minimum, the suitability of MapInfo software for even demanding, highly technical applications is obvious.

CASE STUDY
REAL ESTATE • SITE SELECTION

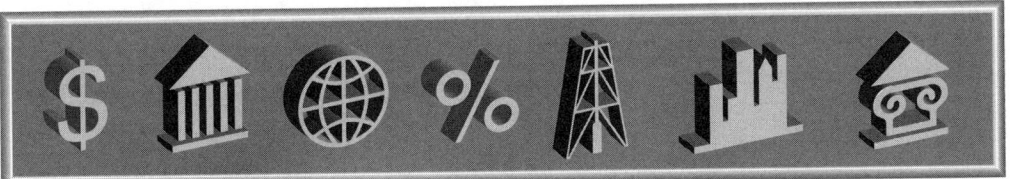

VISIMARK

Visimark LLC was launched in late 1994 by two commercial real estate brokers and an entrepreneur who believed the new mapping technology available through Microsoft and MapInfo could greatly enhance the marketing and management of commercial, corporate, and government real estate. They contracted with a major software development firm to create a user friendly information system that would enable real estate agents, brokers, and managers to automate their practice and selectively share proprietary information.

Versatility

Visimark was designed to eliminate the labor-intensive practice of searching for properties with specific criteria to match client needs, as well as generate presentations and information packages by automating the process. The system stores electronic images of photographs, drawings, and documents in property files and enables agents to maintain proprietary notes. In addition, the system is accessible to multiple users via an electronic central data entry system that maintains database integrity, currency of information, and system security.

Office building interior (left) and lobby (right).

In Visimark, property images are paramount. Used for all types of real estate holdings, the system could display the images below for a client interested in potential manufacturing sites.

Exterior (left) and interior (right) of a manufacturing site.

The process quickly eliminates potential locations so that the broker and client spend far less time conducting on-site tours. In addition, the broker can show specialty locations of virtually any variety.

Searching and Database Capabilities

A sophisticated search engine queries multiple criteria simultaneously to choose appropriate sites. The pictures, drawings, and documents are automatically assembled into a single presentation package. With several clicks of the mouse, the package can be faxed, e-mailed, printed, or downloaded to a laptop for a remote presentation. This efficiency allows clients to tour 15 properties from the office in 20 to 30 minutes, rather than spending several hours driving to each property. In addition, because Visimark is fully compatible with MapInfo, maps containing street data can be generated quickly for any area in the United States and most metropolitan areas in other countries. Demographic data now available in digital form can be developed for each site as well.

Demographic analysis data report for a property in Bannockburn, Illinois.

The Visimark system relies on two separate databases to serve as a multiple listing service for commercial real estate. An internal database is accessible to each broker within a firm, while the Multiple Listing Database (which also accesses Visimark) is a separate database that does not interact with the firm's proprietary information. Moreover, each participating firm can select information it wants to share with the Multiple Listing Database. Agents can maintain contact management software with notes on each client at all times. The software can also be Web enabled to operate on the Internet, an intranet, or an extranet.

Saving Time, Money, and Drive Space

Because images are vital to the system's usefulness, the Visimark imaging engine—which must handle an unlimited number of images—had to be robust. In the system, pictures are accessed quickly, have extraordinarily high resolution, and take up minimal disk space. For example, a 1,200,000 byte (1.2Mb) picture is compressed to 40,000 bytes (40K), saving storage space and enabling the image to be transmitted quickly via electronic means—far faster than over ISDN lines. The ability to print precise and limited quantities of materials saves on printing costs.

On the property and asset management side, the software is ideal for cataloging properties and equipment. Improvements in asset management, maintenance, reporting, construction planning, engineering coordination, and risk management are all feasible. Having instant access to pictures and drawings can resolve conflicts in minutes rather than weeks or days. The system also facilitates consensus decision making and cuts travel time and cost.

GIS Interface

When designing a visual information system for use in real estate, Visimark's developers knew that the product had to take advantage of the latest advances in GIS. Integrating Visimark with the most widely used GIS software, MapInfo Professional, enables end users to analyze and visualize data regarding their assets and properties in a new way.

Developers also confirmed that many target users of the application—especially real estate professionals—had little computer knowledge or experience. Although Visimark had an intuitive, user friendly format, using the GIS package would require more sophistication. The goal was to allow users to take advantage of the power of MapInfo without expensive, time-consuming training.

To achieve this goal, Visimark programmers worked closely with a MapBasic team to develop a complex program that would allow Visimark and MapInfo to communicate with one another. The result, MapLink, has a user friendly interface and full GIS functionality. This means that users can access MapInfo maps of specific sites with surrounding roads and points of interest highlighted. Such functionality illustrates for clients the precise location of commercial properties. For example, the following illustration shows a view of the Jiffy Lube site relative to the Tri-State Tollway.

GIS Interface

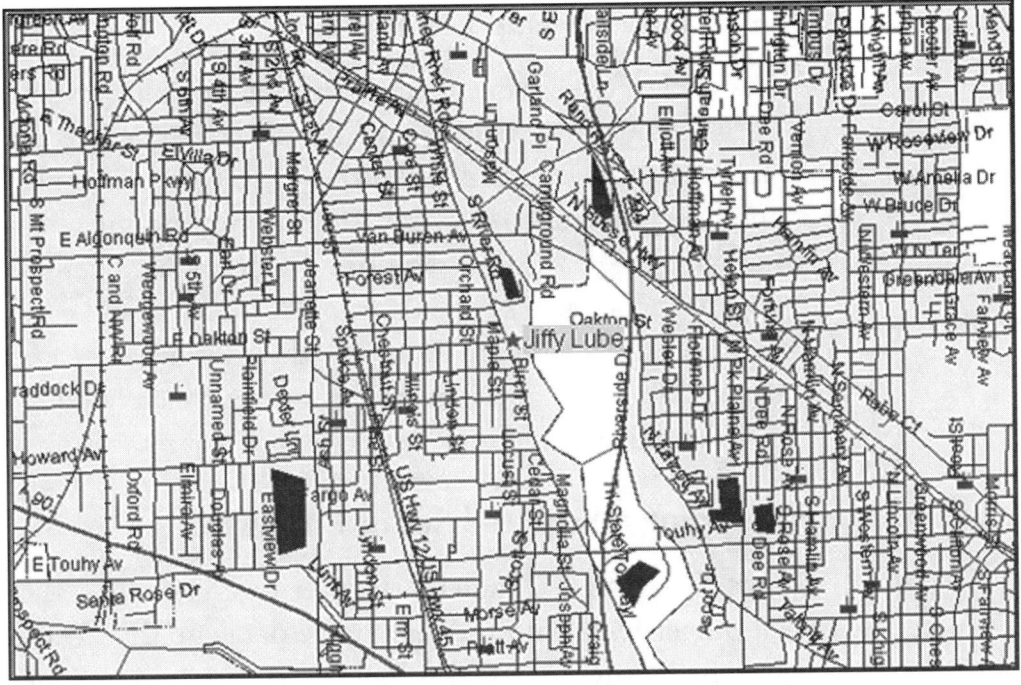

Detailed community map of a Jiffy Lube location.

After reviewing the existing Visimark program and the needs of most real estate professionals, programmers decided to use MapLink to automate three primary GIS functions within Visimark: geocoding, searching by map location, and mapping a selected set of properties. Each function is described below.

Geocoding

Geocoding is the process by which latitude and longitude are assigned to a property, generating a set of coordinates that enable MapInfo to display the property on a map. MapLink allows Visimark to automatically geocode properties once addresses are entered.

Searching by Map Location

It was clear to Visimark's developers that simply typing in a community name would not be sufficient for searches. Instead, MapLink allows users to access a map and select an area to search by simply pointing and clicking with the mouse. In response, MapLink analyzes the selected area, breaks it down into data sectors, and communicates the information to the Visimark search engine. The user, unaware of this process, merely points to a desired area and reviews the results.

Mapping a Selected Set of Properties

Once a search is complete, the user receives a hit list of all properties within the searchable database that fit the specified criteria. At this point, MapLink intervenes again to display an instant "live" map of the properties on the hit list. The user can include other forms of digitized GIS data, such as demographics, for further analysis. MapInfo's unique overlay capability allows countless levels of data to be simultaneously viewed for fast, accurate decision making and analysis.

Using standard GIS map tools such as zooming in, zooming out, and panning, the user is able to position and reposition the map to view the location. The following map illustrates the same Jiffy Lube location as the previous map with a greater zoom. The client is now able to see more of the surrounding street network.

GIS Interface

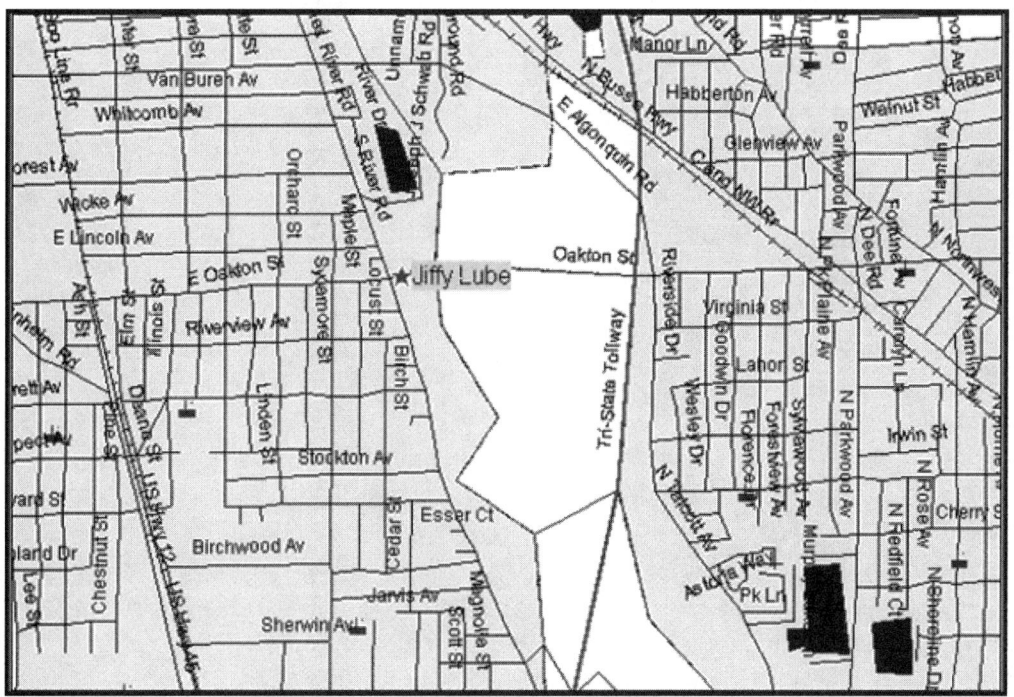

Zooming in provides a closer view.

The resulting integrated product is ideal for developers, asset managers, and real estate professionals.

APPENDIX A

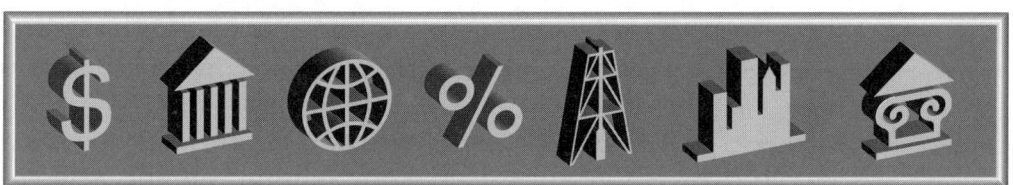

GLOSSARY

Address matching—Assigning an absolute location, through XY coordinates, to each address in a MapInfo table. This process occurs through interpolating specific address locations against a street file with address ranges. See *Geocoding*.

American Standard Code for Information Interchange (ASCII)—A standard code used for internal computer character representations.

Attribute data—Also known as tabular data, attribute data are linked to map objects. The attribute data linked to a particular map object may contain geographic information (e.g., addresses or ZIP Codes). Attribute data can also contain information associated with features in a map object, such as soil properties or land use descriptions.

Base map—A map containing visible surface features and boundaries, accurately referenced to a specific coordinate system.

Browser—A window in MapInfo that allows the user to view and manipulate attribute data in a tabular or spreadsheet format.

Buffer—A region created in MapInfo around selected geography drawn at a specified distance from the object. Buffers are used for various types of proximity analysis (e.g., trade area analysis).

Census block—Usually a small area (e.g., city block) used to compile the population count from periodic census studies performed by the U.S. Census Bureau.

Census block group—An area made up of several *census blocks*.

Centroid—The "center" point of a map object. In MapInfo, the centroid of an object is used for auto labeling, geocoding, and placement of thematic charts.

Coordinate system—A map reference system in which the precise geographic positions comprising a local area can be referenced by means of a rectangular grid. The use of a rectangular grid allows features to be located using XY coordinates. This system facilitates the integration of survey data into a larger national grid.

Dot density thematic map—A map format showing data as small dots. Each dot represents a quantity of data (e.g., 100 households).

Export—The process of moving a file prepared in MapInfo to another application.

Field—A unique descriptor or characteristic of a *record* (or instance) in a database; also know as a column or item. In a customer database, examples of fields could be name, address, city, and ZIP Code.

Geocoding—The process of assigning an absolute location (in XY coordinates) to a geographic feature referenced by a relative location, such as a street address or ZIP Code.

Geographic coordinates—The geographic reference system of latitude and longitude in which the Earth is treated as a sphere and divided into 360 equal parts (or degrees). This division is performed along two axes, one running east-west along the equator, and the other running north-south along the Greenwich Prime Meridian. Using this coordinate system, any location on the Earth can be identified with a unique XY coordinate pair. Geographic coordinates are commonly measured in degrees, minutes, and seconds and can also be formatted as decimal degrees.

Geographic information system (GIS)—A geographic database manager that treats all geographic (or spatial) features as records in a database, not simply as graphics. Virtually all concepts in traditional relational databases apply to GIS, but with the added dimension of geography. A GIS builds a bridge between geography and descriptive information through a georelational model that establishes a one-to-one relationship between a spatial data set and an attribute table. Some database fields are predefined for you in attribute tables, but you can add fields as desired. The georelational model also permits you to connect to other tabular databases, whether internal or external to the GIS software.

Global positioning system (GPS)—A 3D surveying system based on signals broadcast from a constellation of Earth-orbiting satellites. One such system, implemented by the U.S. Department of Defense, is available for civilian use.

Graduated symbols thematic map—A map format showing data as symbols drawn at various sizes.

Import—The process of loading data into MapInfo from an external source. For example, if you import dBase or delimited text files, you are taking files created outside of MapInfo and incorporating (or importing) them into the software.

Layer—A group of features or objects comprising the fundamental elements of a map. A single layer is usually composed of similar objects, such as streets, rivers, or store locations. Multiple layers (or tables of objects) are stacked together to create a map, like a series of transparent overlays.

Layout—A map composition document used to prepare output from MapInfo. A layout allows the user to define and arrange maps, browsers, graphs, and other information for professional looking printed output.

Legend—A box that translates the colors and styles used in a map into data values.

Line—One of the three fundamental GIS object types (point, line, polygon), lines are located and defined through the assignment of a connected series of XY coordinate pairs.

Open database connectivity (ODBC)—A standard that allows remote databases to be "attached" to MapInfo applications. For example, a remote Oracle database can be attached through use of ODBC connections so that you can display an Oracle data table in a MapInfo Browser window.

Pie chart thematic map—A map format displaying small pie charts at each record's map location.

Point object—One of the three fundamental GIS object types (point, line, polygon), points represent entities found at discrete locations, such as oil wells, transformer sites, or customer locations. Each point is located using a single XY coordinate pair.

Polygon—One of the three fundamental GIS object types (point, line, polygon), polygons represent entities of a real extent, such as states, land parcels, geologic zones, or islands. Polygon features are defined by a series of XY coordinate pairs identifying the polygon's perimeter.

Query table—A temporary MapInfo selection table that can be created using one of the selection toolbar items or using the SQL Select dialog box. MapInfo automatically labels query tables "Query1," "Query2," and so forth.

Record—A specific instance or member of a database; a record is also known as a row. In a customer database, records could be the information pertaining to individual customers.

Redistricting—The process of grouping map objects. Redistricting is used for problems such as creating sales districts or maintaining voter precincts.

Routing—Determining the path between multiple points. The path generally follows a route system such as a road network or power lines. For example, a delivery route would be the path on the roads the driver would travel to reach each delivery point.

Scale—The ratio of distance covered on a photograph or map to its corresponding distance on the ground. Scale can be expressed as a ratio (1:10,000), a fraction (1/10,000), or an equivalent with units (1" = 833.33 ft).

Spatial analysis—An operation that examines data with the intent to extract or create new data that fulfill a required condition. Spatial analysis usually involves operations such as "contains," "intersects," or "within" to analyze the relationship between buffers or polygons and another geographic layer.

Spatial data—Data that identify the location of geographic objects and describe their spatial dimensions. Spatial data are classified as points, lines, or polygons.

Structured query language (SQL)—Usually pronounced "sequel," SQL is an official standard relational language for databases approved by an ANSI committee in 1986. There are many implementations of SQL based on the published standard.

Table—A basic unit of storage in a database management system containing a two-dimensional matrix of attribute values (i.e., rows and columns).

Thematic map—A map in which variations in graphic style (e.g., color, pattern, size, and symbol) are used to illustrate differences in the underlying data.

Topologically Integrated Geographic Encoding and Referencing (TIGER)—U.S. Census Bureau reference files developed for the 1990 census. The TIGER system is designed to fully integrate with actual base map data from 1:100,000 DLG files created by the USGS.

XY coordinate—The unique geographic location of a spatial feature. All spatial data are maintained in a specific map coordinate system, the most widely used of which is the geographic coordinate system, latitude and longitude.

Zoom layering—The MapInfo setting for layer visibility. For example, a detailed street layer may be set to be visible only in a map window that displays an area less than 5 mi wide. Zoom layering helps control what layers of data are displayed as you zoom in and out on a map.

Appendix B

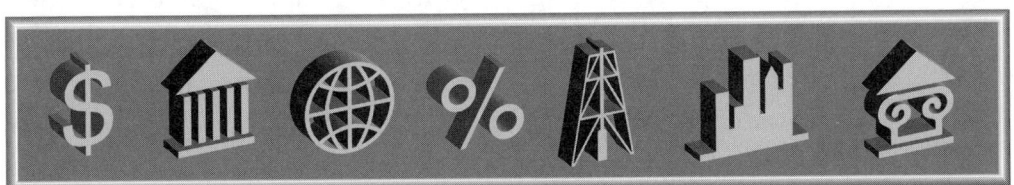

Reference Material and Data Sources

OnWord Press MapInfo Books

Inside MapInfo Professional (1996). A users' tutorial for learning MapInfo's mapping software. Also includes case studies illustrating how several companies use MapInfo.

Mapping with Microsoft Office (1996). Introduces the capabilities of Microsoft Map, the MapInfo-provided mapping feature in Microsoft Excel released with Office 95.

MapBasic Developer's Guide (1996). Trains users to write MapBasic programs. The MapBasic language can be employed to create custom mapping applications or simply add functionality to MapInfo.

E-mail List

To subscribe to an extremely active MapInfo discussion electronic mailing list (it averages approximately 20 messages per day), send an e-mail message to *majordomo@csn.org*. In the subject of the message type *subscribe mapinfo-l*. (This list can also be accessed from MapInfo's Web site at *www.mapinfo.com*.)

Satellite and Aerial Data Sources

Source	Phone number and Web address
American Society for Photogrammetry and Remote Sensing Washington, DC	301-493-0290 http://www.asprs.org/asprs
Autometric Inc. Alexandria, VA	703-658-4000 http://www.autometric.com
Earth Satellite Corp. Rockville, MD	301-231-0660 http://www.earthsat.com
EarthWatch Inc. Longmont, CO	303-682-3800 http://www.digitalglobe.com
EDRAS Inc. Atlanta, GA	444-248-9000 http://www.edras.com
EROS Data Center, U.S. Geological Survey Sioux Falls, SD	605-594-6151 http://www.edcwww.cr.usgs.gov/eros-home.html
OrbImage Dulles, VA	703-406-5436 http://www.orbimage.com

Remote Sensing Resources on the Web

Source	Web address
University of Manchester Manchester, United Kingdom	http://www.man.ac.uk/Arts/geography/rs/rs.html
Space Imaging EOSAT Thornton, CO	303-254-2000 http://www.spaceimage.com
SPOT Image Corp Reston, VA	703-715-3100 http://www.spot.com

Telecom Data

The following is a list of telecom data products available through MapInfo. Contact the company for more information.

MapInfo product	Application					
	Competition	Customer Service	Opportunity Analysis	Engineering	Wireless	Sales & Marketing
AreaCode Info—Maps the boundaries of U.S. three-digit area codes.	X	X	X		X	X
CellularInfo—A mapping database of cellular coverages in every cellular market in the United States.	X	X	X	X	X	X
Exchange-Info—A database containing wire center serving areas.	X					

MapInfo product	Application					
	Competition	Customer Service	Opportunity Analysis	Engineering	Wireless	Sales & Marketing
Exchange-Info Plus— A database of estimated wire center serving areas with variables such as LATA, NPA, NXXs within the serving area; local operating company name, rate center name, and switch information.		X	X	X	X	X
Highway Volumes— Provides 24-hour average daily traffic counts for U.S. interstates, U.S. highways, and state highways.		X	X	X	X	X
LATAInfo—A map database of Local Access and Transport Area (LATA) boundaries.	X	X	X	X	X	X
MSA/RSA Cellular Markets— Metropolitan Service Areas (MSA) and Rural Service Areas (RSA) markets.	X	X	X		X	X

Telecom Data

MapInfo product	Application					
	Competition	Customer Service	Opportunity Analysis	Engineering	Wireless	Sales & Marketing
PCSInfo—A database of U.S. markets and companies awarded each of the A, B, C, D, E, and F block licenses. Includes BTA/MTA boundaries and 1990 census demographic data.	X		X	X	X	
Traffic Volumes—Provides 24-hour average daily traffic counts for highways and major roads in all U.S. metropolitan areas.		X	X	X	X	X

Bureau of Transportation Statistics Data

Electronic data	
Name	**Description**
1990 Census Transportation Planning Package (CTPP)	A specially tailored census database for transportation planners. Data include information about people, workers, and housing units by geographic area. Includes characteristics of workers in journey-to-work flows.
International Travel and Tourism (T&T)	Contains travel and tourism statistics from national and international organizations worldwide. Represents approximately 100 nations.
National Transportation Atlas Databases	Geographic databases that provide the infrastructure for national planning and policy initiatives. Atlas database includes airports, Amtrak and railway stations, water ports, railroad networks, highways, and waterways. Also includes many region boundaries such as states, counties, and national parks.
Rail Waybill data	All non-confidential rail shipment data such as origin and destination regions, type of commodity, number of cars, tons, revenue, length of haul, and interchange states.
TIGER/Line files	Reference files developed by the U.S. Census Bureau for the 1990 census. The TIGER system is designed to fully integrate with actual base map data from the USGS 1:100,000 digital line graph files. The TIGER files are the basis from which most vendors of stored street data build their databases.
Traffic safety data	National information on accident fatality rates collected by the U.S. Department of Transportation's National Highway Traffic Safety Administration. Also have state and metropolitan traffic safety information on over 200 safety related reports, studies, programs, and so forth.

Hardcopy data	
Name	**Description**
Commodity Flow Survey (CFS)	Reports the value, weight, mode, and distance transported of commodities shipped by manufacturing, mining, wholesale, trade, and selected retail and service industries.
Directory of Transportation Data Sources	Provides a comprehensive inventory of transportation data sources to foster accessibility to information.

Hardcopy data	
National Transportation Statistics (NTS)	Historical compendium of selected national transportation and transportation related data from a wide variety of government and private sources. Includes safety and motor vehicle sales, as well as energy statistics (e.g., consumption).
Worldwide Transportation Directory	Lists 1,925 transportation profiles and contact points in 189 countries. Data entries are primarily government and quasi-governmental agencies and organizations.
NOTE: For contact information, see Appendix D, "Contact Information."	

Internet Sites

A special thanks to Dr. Dennis Fitzsimons, who created the following list of interesting GIS-related Internet sites.

Agencies

Source	Web address
1990 U.S. Census LOOKUP	http://cedr.lbl.gov/cdrom/doc/lookup_doc.html
ACSM Home Page	http://www.landsurveyor.com/acsm/
Bureau of the Census	http://www.census.gov/
Central Intelligence Agency	http://www.odci.gov/cia/
ERIN	http://kaos.erin.gov.au/erin.html
Federal Geographic Data Committee (FGDC)	http://www.fgdc.gov/
FEMA	http://www.fema.gov/homepage.shtml
Government Publications on the Web	http://www.library.nwu.edu/gpo/
NASA Homepage	http://www.gsfc.nasa.gov/NASA_homepage.html
National Geodetic Survey	http://www.ngs.noaa.gov/index.html
Natural Resources Canada (NRCan)	http://www.nrcan.gc.ca/home/nrcanhpe.htm
NIMA homepage	http://164.214.2.59/nimahome.html
NOAA Coastal & Estuarine	http://www-ceob.nos.noaa.gov/
State and Local Government	http://www.piperinfo.com/piper/state/states.html
Federal Government Agencies	http://www.lib.lsu.edu/gov/fedgov.html
U.S. Fish and Wildlife Service	http://www.fws.gov/
U.S. Geological Survey	http://www.usgs.gov/

Source	Web address
USGS Water Resources of the United States	http://h2o.usgs.gov/
USGS Mapping Information	http://www-nmd.usgs.gov/
USGS Publications Index Guide	http://andriot.com/USGS.htm
World Health Organization	http://www.who.ch/

Cartography

Source	Web address
1992 National Resources Inventory Atlas	http://www.nhq.nrcs.usda.gov/nriatlas.html
Abbreviations & Acronyms	http://www.lib.berkeley.edu/EART/abbrev.html
ACMLA (Association of Canadian Map Libraries)	http://www.sscl.uwo.ca/assoc/acml/acmla.html
Alexandria Digital Library	http://alexandria.sdc.ucsb.edu/
Baltic GIS Database	http://www.grida.no/baltic/
Barcelona Map Exhibit	http://www-nais.ccm.emr.ca/barcelona_map_exhibit/estart.htm
Berkeley Map Collection	http://www.lib.berkeley.edu/EART/digital/tour.html
Boston Transport Services	http://www.mbta.com/~imagemap/GIFBAR?139,5
Bowen U.S. maps	http://130.166.124.2/USpage1.html
California Atlas (Bowen)	http://130.166.124.2/CApage1.html
Cartographic/Geographic Web Sources – University of Georgia	http://www.libs.uga.edu/maproom/ahtml/mchpcr1.html
Cartographic Communication	http://www.utexas.edu/depts/grg/gcraft/notes/cartocom/toc.html#3.3
Cartographic Glossary	http://www.lib.utexas.edu/Libs/PCL/Map_collection/glossary.html
Cartographic Images – Siebold	http://www.iag.net/~jsiebold/carto.html
Cartographic Materials – University of Waterloo	http://www.lib.uwaterloo.ca/discipline/Cartography/cart.html
Cartographic Records (Digital) – Pilot	http://www.bcars.gs.gov.bc.ca/cartogr/general/maps.html
Cartographic Reference Books – Philadelphia Print Shop	http://www.philaprintshop.com/cartrftx.html
Cartographic References – PCL	http://www.lib.utexas.edu/Libs/PCL/refserv/geography/Cartographic_reference.html
Cartography – Calendar of Events	http://www.cyberia.com/pages/jdocktor/
Cartography – Indiana State	http://www.indstate.edu/gga/gga_cart/index.html
Cartography Info Center – LSU	http://www.cadgis.lsu.edu:80/cic/

Internet Sites

Source	Web address
Cartography Resources – GMU	http://geog.gmu.edu/gess/jwc/cartogrefs.html
Census Index of /mapGallery/images/	http://www.census.gov/ftp/pub/geo/www/mapGallery/images/
Centennia	http://www.clockwk.com/
CGRER NetSurfing: Maps and References	http://www.cgrer.uiowa.edu/servers/servers_references.html
Chicago 1990 Census Maps	http://www.lib.uchicago.edu:80/LibInfo/Libraries/Maps/chimaps.html
Color Landform Atlas – U.S.	http://fermi.jhuapl.edu/states/states.html
Color Use Guidelines	http://www.gis.psu.edu/Brewer/CBColorHTML/CBColorTop.html
Country maps from W3 servers in Europe	http://www.tue.nl/europe/
Digital Chart of the World	http://ilm425.nlh.no/gis/dcw/dcw.html
Digital Wisdom	http://www.digiwis.com/
DMSP City Lights	http://web.ngdc.noaa.gov/dmsp/IMAGERY/newols-app-city.html
Election results (Clinton)	http://www.geog.ucsb.edu/~lawson/election.html
Encyberpedia's MAPS and geography	http://www.montecristo.com:80/map1.htm
GeoSystems: Map Skills	http://www.geosys.com/cgi-bin/genobject/mapskills/tigdd27
GMU Geo 310 Maps	http://geog.gmu.edu/gess/jwc/student_projects.html
How far is it?	http://gs213.sp.cs.cmu.edu/prog/dist
IASBS Digital Map Files	http://www.usm.maine.edu/~maps/iasbs/digmaps.html
ICA: U.S. National Commission	http://www.gis.psu.edu/ica/ICAusnc.html
Imaginary Maps	http://www.gsi-mc.go.jp/tizu/animap.html
Java Atlas Home page	http://www.ggr.ulaval.ca/JAVA/Java.html
Links to Maps	http://www.cco.caltech.edu/~salmon/maps.html
London Maps	http://multimap.com/london/
Map Images on the Web	http://www.cadgis.lsu.edu:80/cic/mapsnet.html
Map Links	http://www.lbl.gov/Web/Maps.html
Map Room – Oxford	http://www.bodley.ox.ac.uk/users/nnj/
MapArt Gallery	http://www.map-art.com/map-art/software/gallery.html
MapFinder	http://fieber-john.campusview.indiana.edu/mapfinder/
MapLink: Online Directory	http://www.maplink.com/Mldir1.htm
Mapmaker	http://loki.ur.utk.edu/ut2kids/maps/map.html

Source	Web address
Maps & GIS INFOMINE Search Screen	http://logic17.ucr.edu//mapsinfo.html
Maps – Nottingham	http://acorn.educ.nottingham.ac.uk/ShellCent/maps/welcome.html
Mapville Home Page	http://www.mapville.com/
Multi-scale Maps	http://www.c3.lanl.gov/~cjhamil/browse/main.html
National Atlas Information – Canada	http://www-nais.ccm.emr.ca/naisgis.html
New York City Maps	http://www.soc.qc.edu/Maps/
New York Map Portfolio	http://www.sunysb.edu/libmap/nymaps.htm
North American Breeding Birds – Ranges	http://www.npsc.nbs.gov:80/resource/distr/birds/breedrng/breedrng.htm
NSW SoE 1995 - Maps	http://www.epa.nsw.gov.au/soe/95/listmaps.htm
NYC Subway Map Picker	http://www.mediabridge.com/nyc/transportation/subways/picker.html
Oddens's Bookmarks	http://kartoserver.frw.ruu.nl/html/staff/oddens/oddens.htm
Oversized Color Maps	http://www.cc.columbia.edu/imaging/html/largemaps/oversized.html
PCL Map Collection	http://www.lib.utexas.edu/Libs/PCL/Map_collection/Map_collection.html
Pilot – Cartographic Records	http://www.bcars.gs.gov.bc.ca/cartogr/general/maps.html
Presidential Elections in Maps	http://www.lib.virginia.edu/gic/elections/index.html
Project Argus – Visualization	http://severn.geog.le.ac.uk/argus/
Road Map Collectors of America	http://falcon.cc.ukans.edu/~dschul/rmca/rmca.html
Shand – Map-related Web Sites	http://www.geog.gla.ac.uk/sites/mapsites.htm
Southern California Area Maps	http://artscenecal.com/Maps.html
Univeristy of California-San Diego – Map Room	http://gort.ucsd.edu/mw/maps.html
U.S. Thematic Maps	http://oseda.missouri.edu:80/graphics/us/pop/
University Campus Maps	http://www.lib.uwaterloo.ca/discipline/Cartography/campus.html
Useful Sites about Maps	http://research.umbc.edu/~roswell/mipage.html
Vinland Map	http://www.mcri.org/vm_image.html
Vinland Map and Shroud Updates	http://www.mcri.org/vm_shroud_update.html#anchor544696
Vinland Map – largest photo	http://www.digalog.com/viking/vinland/m/l/vmap.htm
Yahoo Maps	http://www.yahoo.com/Science/Geography/Maps

Commercial

Source	Web address
Adobe Systems	http://www.adobe.com/
adobe.mag	http://www.adobemag.com/
Adventurous Traveler Bookstore – Map List	http://www.gorp.com/atb/maps.htm
Agfa Home	http://www.agfa.com/
Apple Computer Technical Support	http://www.support.apple.com/
Argus Technologies Homepage	http://www.argusmap.com/
Avenza Software Marketing Inc. MAP-MAC Page	http://www.avenza.com/map-mac.html
Buydirect.com	http://www.buydirect.com/
Cartesia	http://www.map-art.com/
Commercial Geography Resources	http://lorax.geog.scarolina.edu/geogdocs/otherdocs/comm.html
Computer Companies Information Center	http://library.microsoft.com/compcos.htm
DeLorme	http://www.delorme.com/home.htm
Environmental Systems Research Institute, Inc.	http://www.esri.com/
Etak Incorporated	http://www.etak.com/
Fodor's	http://www.fodors.com/
Forefront-eduspecial	http://www.ffg.com/edu/eduspecial.html
Four One Company Ld	http://www.icis.on.ca/fourone/
GIS World, Inc.	http://www.gisworld.com/index.html
Government Technology	http://www.govtech.net/
Intergraph	http://www.intergraph.com/
Mac Zone Home	http://www.maczone.com/maczone?mzstart@55884jhkm
MacAddict	http://www.macaddict.com/
MacWarehouse	http://www.warehouse.com/MacWarehouse/
Macworld Online Buyers Guide	http://www.macworld.com/buyers/hot.deals/index.html
MAGELLAN Geographix	http://www.magellangeo.com/
MapInfo Corporation	http://www.mapinfo.com/homepage.html
Mercator's World Home Page	http://www.mercatormag.com/hihome.htm
Mountain High Maps	http://www.digiwis.com/
OMNI Resources (topographic sheets)	http://www.omnimap.com/catalog/index.htm
Power Computing Corporation	http://www.powercc.com/

Source	Web address
Rand McNally	http://www.randmcnally.com/home/
Raven Maps & Images	http://www.ravenmaps.com/
Silicon Graphics' Silicon Surf	http://www.sgi.com/
Simm.Net	http://194.72.252.2:80/simmnet/
Sirs, Inc.	http://www.sirs.com/
Steven Gordon Cartography	http://www.clearwater.com/gordonmaps/
Sure!MAPS Digital Mapping	http://www.horizons.com/suremaps
Terra Data/Geocart	http://hudson.idt.net/~terrainc/
The Gold Bug	http://www.goldbug.com/
ThinkSpace/Map Factory	http://www.thinkspace.com/
United Computer Exchange Corporation	http://www.uce.com/
World of Maps	http://www.WorldofMaps.com/

Data Sources

Source	Web address
1990 U.S. Census LOOKUP	http://cedr.lbl.gov/cdrom/doc/lookup_doc.html
Assoc. of Research Librarians Statistics and Information	http://www.lib.virginia.edu/socsci/arl/test-arl/index.html
Bay Area Regional Database	http://bard.wr.usgs.gov/
Bureau of the Census	http://www.census.gov/
Bureau of Transportation Statistics	http://www.bts.gov/btsprod/order.html
CGRER NetSurfing: Maps and References	http://www.cgrer.uiowa.edu/servers/servers_references.html#interact-generalCIE
SIN Home Page	http://www.ciesin.org/
CIESIN: Directory /pub/census	ftp://ftp.ciesin.org/pub/census/
Demography and population	http://coombs.anu.edu.au/ResFacilities/DemographyPage.html
ENRM prototype server	http://enrm.ceo.org/home.pl
EROS Data Center	http://edcwww.cr.usgs.gov/doc/edchome/datasets/edcdata.html
FGDC Subcommittee on Cultural and Demographic Data	file://www.census.gov/pub/geo/www/standards/scdd/index.html
Global Land Information System	http://edcwww.cr.usgs.gov/webglis/
Government Information Sharing Project	http://govinfo.kerr.orst.edu/index.html

Internet Sites

Source	Web address
GPO-Federal Locator Services	http://www.access.gpo.gov/su_docs/dpos/adpos400.html
Mable/Geocorr Home Page	http://ts2.ciesin.org/plue/geocorr/
Mable/Geocorr 2.01 Home Page	http://www.oseda.missouri.edu/plue/geocorr/
National Geospatial Data	http://nsdi.usgs.gov/nsdi/
Northern Prairie Science Center	http://www.npsc.nbs.gov/
Office of Social & Economic Data Analysis (OSEDA)	http://www.oseda.missouri.edu/
Ohio State University Spatial Data Sources	http://ncl.sbs.ohio-state.edu/5_sdata.html
Population Reference Bureau	http://www.prb.org/prb/index.html
PSU-Earth and Mineral Sciences Library	http://vector.gis.psu.edu/emsltop.html
Regional Economic Information System, 1969-1993	http://www.lib.virginia.edu/socsci/reis/reis1.html
SEDAC-Socioeconomic Data	http://sedac.ciesin.org/
U.S. Census Data-Lawrence Berkeley National Lab	http://cedr.lbl.gov/mdocs/LBL_census.html
University of Arkansas-Japan GIS/Mapping Sciences Resource Guide	http://www.cast.uark.edu/jpgis/
University of Connecticut Map Library-MAGIC	http://magic.lib.uconn.edu/
UN gopher	gopher://gopher.undp.org:70/11/ungophers/popin
USA Counties 1996	http://govinfo.kerr.orst.edu/usaco-stateis.html
USGS NSDI - DEMs	http://nsdi.usgs.gov/nsdi/products/dem.html
World Factbook Master Home Page - CIA	http://www.odci.gov/cia/publications/nsolo/wfb-all.htm

Earth Sciences

Source	Web address
ABAG Earthquake Maps and Information	http://www.abag.ca.gov/bayarea/eqmaps/eqmaps.html
Caltech: Frequently-used Resources	http://www.caltech.edu/caltech/Frequent.html
Central Michigan University Geography and Earth Science	http://www.cmich.edu/~3nrwbhg/homepage.htm
CIESIN Gateway WWW Interface	http://wwwgateway.ciesin.org/
CSC Earth Science Topics	http://www.csc.fi/earth_science/earth_science.html
Earth Pages	http://starsky.hitc.com/earth/earth.html
Earth Viewer	http://fourmilab.ch/earthview/vplanet.html
Earth2 Project	http://www.ems.psu.edu/Earth2/E2Top.html

Source	Web address
Earthmap Home Page	http://www.gnet.org/earthmap/
Earthquake Information (USGS)	http://quake.wr.usgs.gov/
EarthView	http://www.ldeo.columbia.edu/EV/EarthViewHome.html
EdWeb Home Page	http://edweb.cnidr.org:90/
Eisenhower National Clearinghouse	http://enc.org/
GAIA Alert	http://www.newciv.org/millennium_matters/gaia
Gap Analysis Home Page	http://www.gap.uidaho.edu/gap/index.html
Geosciences Resources	http://www.cc.columbia.edu/cu/libraries/indiv/geosci/offsite.html
GeoWeb Home Page	http://wings.buffalo.edu/geoweb/
Gisnet - Online Resources for Earth Scientists	http://www.gisnet.com/gis/ores/gis/hyper.html
McKnight Test Pages	http://www.prenhall.com/divisions/esm/mcknight/public_html/
Meteorology Obs: References	http://www.ems.psu.edu/~fraser/Meteo471/Meteo471wRefMeteo.html
NASA – EOS Project Science Office	http://eospso.gsfc.nasa.gov/
Northern Prairie Science Center	http://164.159.215.66/
NOSC - Planet Earth home page	http://www.nosc.mil/planet_earth/everything.html
Physical Geography Resources	http://feature.geography.wisc.edu/phys.htm
PSU – Earth and Mineral Sciences Library	http://vector.gis.psu.edu/emsltop.html
Sirs, Inc.	http://www.sirs.com/
Tasa Exchange Earth Science Links	http://www.swcp.com/~tasa/links.html
Tasa Graphics Earth Exchange	http://www.swcp.com/~tasa/
The GeoSphere Project Report	http://www.infolane.com/geosphere/
The NASA/JPL Imaging Radar Home Page	http://southport.jpl.nasa.gov/
The World Lecture Hall	http://www.utexas.edu/world/lecture/
University of California-Berkeley Internet Resources in the Earth Sciences	http://www.lib.berkeley.edu/EART/EarthLinks.html
UCI Science Education Programs Office	http://www-sci.lib.uci.edu/SEP/SEP.html
UIUC – Online Guide to Meteorology	http://covis.atmos.uiuc.edu/guide/guide.html
USGS – Earth and Environmental Science	http://www.usgs.gov/network/science/earth/earth.html
USGS/Pasadena Home Page	http://www-socal.wr.usgs.gov/
USRA – Earth System Science Education	http://www.usra.edu/esse/ESSE.html
Views of the Solar System	http://bang.lanl.gov/solarsys/

Source	Web address
Virtual Library – Earth Sciences Resources	http://www-vl-es.geo.ucalgary.ca/VL/html/es-resources.html
W.M. Keck Foundation Seismological Observatory Earthquake Information	http://www.baylor.edu/~Geology/keck_eq.html
WebEarth	http://www.hyperreal.com/~mpesce/we/

Environment

Source	Web address
Environmental Organization WebDirectory	http://www.webdirectory.com/
Global Information Locator	http://www.g7.fed.us/gils.html
Greenpeace International Home Page	http://www.greenpeace.org/
IGC: EcoNet	http://www.igc.org/igc/econet/index.html
National Audubon Society	http://www.audubon.org/
National Wildlife Federation	http://www.nwf.org/
Nature Conservancy	http://www.tnc.org/
Rainforest Action Network Home Page	http://www.igc.apc.org/ran/
Sierra Club Home Page	http://www.sierraclub.org/

Geography

Source	Web address
Geography and related data resources	http://www.thomas.com/othergeo.html
Geography and GIS Resources	http://www.clark.net/pub/lschank/web/geo.html
MiSU - Geography-related Servers	http://www.ssc.msu.edu/~geo/geolinks.html
National Geographic Society Online	http://www.nationalgeographic.com/
The Microstate Network	http://microstate.com/cgi-win/mstatead.exe/listregions
UB GIAL – Internet Geography Information	http://www.geog.buffalo.edu/GIAL/netgeog.html
UCR – INFOMINE	http://lib-www.ucr.edu/
University of Texas-Austin Geography Resource Center	http://www.utexas.edu/depts/grg/virtdept/resources/contents.htm
Virtual Library: Geography	http://www.icomos.org/WWW_VL_Geography.html

GIS/Remote Sensing

Source	Web address
AGI GIS Dictionary – Free Edition	http://www.geo.ed.ac.uk/agidict/welcome.html
anaglyphs	ftp://geog.ucsb.edu/pub/tmp/
Arizona Geographic Information Council	http://www.state.az.us/gis3/agic/agichome.html
Autometric, Inc. Home Page	http://www.autometric.com/
Baltic GIS Database	http://www.grida.no/baltic/
CALMIT Nebraska-Lincoln	http://www.calmit.unl.edu/calmit.html
Declassified Satellite Photos	http://edcwww.cr.usgs.gov/Webglis/glisbin/search.pl?DISP
Digital Land Systems Research	http://www.mira.net.au/dlsr/
Earth Observation Magazine	http://www.eomonline.com/
EarthRISE	http://earthrise.sdsc.edu/earthrise/maps/
Erdas	http://www.erdas.com/
EROS home page	http://edcwww.cr.usgs.gov/eros-home.html
EUROGI Homepage	http://www.frw.ruu.nl/eurogi/eurogi.html
GeoWeb For GIS/GPS/RS	http://www.ggrweb.com/
GIS and GIS-Related Net Sites	http://www.hdm.com/gis3.htm
GIS FTP Resource List	http://www.geo.ed.ac.uk/home/gisftp.html
GIS Gopher Resource List	http://www.geo.ed.ac.uk/home/gisgopher.html
GIS/Cartography (Norway)	http://www.iko.unit.no/gis/gisen.html
GIS/Mapping Sciences Resource Guide	http://www.cast.uark.edu/jpgis/jpgsittmf.html
GISnet BBS' MapInfo Support Page	http://www.csn.net/gis/mapinfo/
GPS – Peter Dana (University of Texas-Austin)	http://www.utexas.edu/depts/grg/gcraft/notes/gps/gps.html
GPS Navigation	http://www.inmet.com/~pwt/gps_gen.htm
GPS World Home Page	http://www.gpsworld.com/
Guide to DEMs	http://www.truflite.com/text/demguide.htm
ImageNet	http://www.coresw.com/
MIT + MassGIS Digital Orthophoto Project	http://ortho.mit.edu/
NASA – JSC Digital Image Collection	http://images.jsc.nasa.gov/html/home.htm
NCGIA	http://www.ncgia.ucsb.edu/
REGIS – Environmental Planning/GIS at Berkeley	http://www.regis.berkeley.edu/
Remote Sensing and GIS Information	http://www.gis.umn.edu/rsgisinfo/rsgis.html
RSL World Wide Web (WWW)...	http://www.gis.umn.edu/

Source	Web address
Space Imaging... 1-Meter Satellite Imagery	http://www.spaceimage.com/
The DEM Reader Page	http://www.electriciti.com/~brianw/DEM_Reader.html
The Kingston Centre for GIS	http://giswww.kingston.ac.uk/menu.html
Thoen's Web	http://www.gisnet.com/gis/index.html
University of Edinburgh GIS WWW Resource List	http://www.geo.ed.ac.uk/home/giswww.html
Virtual Library: Remote Sensing	http://www.vtt.fi/aut/ava/rs/virtual/other.html

Historical Cartography

Source	Web address
A G S Collection	http://leardo.lib.uwm.edu/
Appalachian Arts Maps	http://www.athens.net/~aarts/
Behaim Globus	http://www.ipf.tuwien.ac.at/veroeffentlichungen/ld_p_ch96.html
Brock University Map Library Home Page	http://www.brocku.ca/maplibrary/
Carta Historica	http://www.jyu.fi/tdk/hum/historia/carta/index.html
Cartographic Arts	http://www.dogstar.com/carto
Exploring the West from Monticello: Home	http://www.lib.virginia.edu/exhibits/lewis_clark/home.html
FINFO: Finland 500 years on the Map of Europe	http://www.vn.fi/vn/um/mapseng.html
Harry Ransom Humanities Research Center	http://www.utexas.edu/depts/grg/classes/grg374/resource/hrcmaps/maps/hrcmaps.html
Harvard Map Collection's Home Page	http://icg.harvard.edu/~maps/
Heritage Map Museum	http://www.carto.com/
Historic maps of the Netherlands	http://grid.let.rug.nl/~welling/maps/maps.html
Historica	http://www.zynet.co.uk:8001/beacon/html/livhis.html
HIstorical maps (amateur)	http://maps.linex.com/map.html
History of Cartography	http://ihr.sas.ac.uk/maps/mapsmnu.html
International Map Trade Association	http://www.maptrade.org/
Jim Seibold Home Page	http://www.iag.net/~jsiebold/carto.html
Map Libraries – Buffalo	http://wings.buffalo.edu/libraries/units/sel/collections/maproom.html
Map sellers	http://192.160.127.232/cgi-bin/category.pl?=Cartography
MapHist Discussion Group	http://kartoserver.frw.ruu.nl/HTML/STAFF/krogt/maphist.htm

Source	Web address
MapHist Indexed Hard Copies	http://kartoserver.frw.ruu.nl/HTML/STAFF/krogt/maph_hc.htm
Maps at Duke University	http://www.lib.duke.edu/pdmt/maps.html
Matthew H. Edney's Links	http://www.usm.maine.edu/~maps/edney/links.html
Menotomy Maps Video	http://users.aol.com/videomap/disp/video.htm
New York City Library Map Division	http://www.nypl.org/reearch/chss/map/map.html
Osher Map Library	http://www.usm.maine.edu/~maps/oml/
Paulus Swaen	http://www.swaen.com/
Perseus Atlas Project	http://perseus.holycross.edu/PAP/Atlas_project.html
Robert Ross & Co.	http://www.abaa-booknet.com/usa/ross/
Rohrbach Library Map Collection	http://www.kutztown.edu/library/maps/map.html
RYHINER-Project at the University Library of Berne	http://www.stub.unibe.ch/stub/ryhiner/ryhiner.html
Steve Bartrick Antique Prints & Maps – Early British maps	http://www.antiquarian.com/bartrick/prints/rare_uk_maps.html
The Cartographic Creation of New England	http://www.usm.maine.edu/~maps/exhibit2/
The Map Case – Oxford, U.K.	http://www.bodley.ox.ac.uk/guides/maps/mapcase.htm
UMN - Borchart - Map Libraries	http://www-map.lib.umn.edu/map_libraries.html
University of Georgia Map Collection	http://www.libs.uga.edu/maproom/ahtml/mchpi1.html
University of Georgia Rare Map Collection	http://www.libs.uga.edu/darchive/hargrett/maps/maps.html
WebMuseum: Map	http://www.oir.ucf.edu/wm/map/
Yale Map Collection	http://www.library.yale.edu/MapColl/front.htm

Interactive Cartography

Source	Web address
BWCAW Internet Mapserver	http://www.gis.umn.edu/bwcaw/mapping/mapping.html
CDF Map Making Facility	http://spp-www.cdf.ca.gov/mapmaker/
Census Data Access Tools	http://www.census.gov/ftp/pub/main/www/access.html
Census Map Statistics	http://www.census.gov/datamap/www/index.html
Census Thematic Mapping System	http://www.census.gov/themapit/www/
CIESIN DDViewer	http://sedac.ciesin.org/plue/ddviewer/htmls/whtst.html
CIESIN-SEDAC's DDCartogram	http://plue.sedac.ciesin.org/plue/ddcarto/
CIESIN: Access to U.S. Demographic Data	http://www.ciesin.org:2222/nii.html

Internet Sites

Source	Web address
CIESIN: Index of /	http://www.ciesin.org:2222/
Clickable State Maps	http://govinfo.kerr.orst.edu/pub/map.html
Coastline Extractor	http://crusty.er.usgs.gov/coast/getcoast.html
Design Map – Outlines	http://life.csu.edu.au/cgi-bin/gis/Map
DOOGIS: A Dynamic GIS Interface	http://doogis.dis.anl.gov/
ETOPO-5 Map Generator	http://www.evl.uic.edu/pape/vrml/etopo/
Flagstaff USGS Shaded Relief Maps	http://wwwflag.wr.usgs.gov/USGSFlag/Data/shadedRel.html
Great Lakes Map Server	http://epawww.ciesin.org/arc/map-home.html
ICE MAPS: Interactive Calif Envir Mgt	http://ice.ucdavis.edu/ice_maps/
Interactive Mapper For Arkansas	http://www.cast.uark.edu/products/MAPPER/
Interactive Spatial Data Browser (DLG Data) –UVA Library	http://www.lib.virginia.edu/gic/spatial/dlg.browse.html
Interactive species mapping (ERIN)	http://www.erin.gov.au/database/WWW-Fall94/species_paper.html
Map-It – A GMT3 Map Generator	http://crusty.er.usgs.gov:80/mapit/
Menotomy MVideo Page	http://users.aol.com/videomap/disp/video.htm
NAIS Map Home Page	http://ellesmere.ccm.emr.ca/naismap/naismap.html
Oakland Map Room	http://199.35.5.101/index1.htm
Online Map Creation	http://www.aquarius.geomar.de/omc/
Pennsylvania Statistics by County	http://www.maproom.psu.edu/cbp/
Rapid Imaging Software's Home Page	http://www.landform.com/
The Profile Maker v1.0	http://www.geo.cornell.edu/geology/me_na/profile_maker/profile_maker.htmlTiger
Map Service	http://tiger.census.gov/cgi-bin/mapbrowse
TMS Experimental Browser	http://tiger.census.gov/cgi-bin/mapbrowse
TOPO! Interactive Maps	http://www.topo.com/
TruFlite's 3D World	http://www.truflite.com/
Up-to-the-minute Southern California Earthquake Map	http://www.crustal.ucsb.edu/scec/webquakes/
USGS Elevation Database	http://wxp.atms.purdue.edu/usgs.html
Virginia County Interactive Mapper Map Creation Form	http://www.lib.virginia.edu/gic/mapper/tigermap.html
Zip2 Interactive Map	http://www.zip2.com/Scripts/map.dll?java=yest&type=htm&usamap.x=1&searching=yes

Interesting

Source	Web address
DEOS Altimetry Atlas	http://dutlru8.lr.tudelft.nl/altim/atlas/
Earth and Moon Viewer	http://www.fourmilab.ch/earthview/vplanet.html
Interactive Connecticut Map	http://www.cs.yale.edu/HTML/YALE/MAPS/connecticut.html
Metro Denver Temporal GIS Project	http://www.sni.net/~castagne/index.html
UVA Library Geographic Information Center Homepage	http://www.lib.virginia.edu/gic/

Locators

Source	Web address
BigBook: Map Search	http://www.bigbook.com/showpage.cgi/1b2dda27-1-0-0?page=navigator_map
City.Net Maps	http://city.net/indexes/top_maps.html
CitySearch: U.S. Map	http://www.citysearch.com/
DeLorme: CyberRouter	http://route.delorme.com/
EtakGuide	http://www.etakguide.com/
GeoCities	http://www.proximus.com/geocities/
Infoseek	http://www.infoseek.com/Facts?pg=maps.html&sv=N3
MapBlast	http://www.mapblast.com/
MapQuest	http://www.mapquest.com/
Maps On Us	http://www.mapsonus.com/
Power Search (ATMs)	http://visa.infonow.net/powersearch.html
U.S. Gazetteer	http://www.census.gov/cgi-bin/gazetteer
US Gazetteer SUNY-Buffalo	http://wings2.buffalo.edu/cgi-bin/gazetteer
Vicinity Services	http://www.vicinity.com/vicinity/services.html
Virtual Map - Dynamic Street Maps	http://www.virtualmap.com/
Xearth HTML Front End	http://www-bprc.mps.ohio-state.edu/xearth/xearth.html
Xerox PARC Map Viewer	http://pubweb.parc.xerox.com/map
Yahoo! Maps	http://maps.yahoo.com/yahoo/

Map Projections

Source	Web address
Great Globe Gallery	http://hum.amu.edu.pl/~zbzw/glob/glob1.htm
John Snyder's Map Projection	http://orc.dev.oclc.org:9000/LOGIN:entityClearLimits=1:entityChooseDb=1:sessionid=0:dbname=mapbib:next=html/mapbib_simple_search.html
Map Projection Home Page	http://everest.hunter.cuny.edu/mp/index.html
Map Projection Overview - Peter Dana (UT-Austin)	http://www.utexas.edu/depts/grg/gcraft/notes/mapproj/mapproj.html
Peters Projection	http://www.webcom.com/~bright/petermap.html

Search

Source	Web address
All4One Search Engines	http://www.all4one.com/
All-in-One Search Page	http://www.albany.net/allinone/
AltaVista Technology	http://www.altavista.digital.com/
Argus Clearinghouse	http://www.clearinghouse.net/
AT&T Internet Toll Free 800 Directory	http://www.tollfree.att.net/dir800/
Bartlett Familiar Quotations	http://www.columbia.edu/~svl2/bartlett/
BigBook Directory Search	http://www.bigbook.com/
Bigfoot Home Page	http://bigfoot.com/
British Library	http://www.bl.uk/
CIESIN - Information Resources	http://www.ciesin.org/home-page/library.html
CIESIN Gateway	http://wwwgateway.ciesin.org/
City.Net	http://www.city.net/
Cosmix Mother Load - Insane Search	http://www.cosmix.com/motherload/insane/
Crayon	http://sun.bucknell.edu/~boulter/crayon/
E-Zines Database Query	http://www.dominis.com/Zines/query.shtml
Educational Hotlists	http://sln.fi.edu/tfi/hotlists/hotlists.html
ElNet Galaxy	http://galaxy.einet.net/galaxy.html
Excite Netsearch	http://www.excite.com/
Four11 White Page Directory	http://www.four11.com/
Genealogy Resource	http://godzilla.westworld.com:80/~mgunn/

Source	Web address
GEONAME	http://www.gdesystems.com/IIS/SlipSheets/GEONAME.html
GNN Home Page	http://www.gnn.com/
HotBot	http://www5.hotbot.com:5555/
Hotlist: Geography	http://sln.fi.edu/tfi/hotlists/geography.html
InfoSeek	http://www.infoseek.com/
Infoseek Ultra	http://ultra.infoseek.com/
Jumbo	http://www.jumbo.com/
Libraries on the Web	http://chehalis.lib.washington.edu/libweb/usa.html
LookUP!	http://www.lookup.com/
Lycos, Inc. Home Page	http://lycos.cs.cmu.edu/
MetaCrawler Searching	http://metacrawler.cs.washington.edu:8080/index.html
Movie Database	http://www.msstate.edu/Movies/search.html
NAISMap Home Page	http://ellesmere.ccm.emr.ca/naismap/naismap.html
Nice Geography/GIS Servers	http://www.frw.ruu.nl/nicegeo.html
OKRA	http://okra.ucr.edu/okra/
Open Text	http://www.opentext.com/
ProFusion	http://www.designlab.ukans.edu/ProFusion.html
SavvySearch	http://guaraldi.cs.colostate.edu:2000/./
Search InfoMac Archives	http://www.netam.net/~baron/infomac/
Search Univ. of Michigan Software Archives	http://www.netam.net/~baron/umich/
search.com	http://www.search.com/
Starting Point	http://www.stpt.com/
SUNY Buffalo Libraries	http://wings.buffalo.edu/libraries/units/sel/electronic.html
Switchboard	http://www.switchboard.com/
Virtual Library: Subject Catalogue	http://www.w3.org/pub/DataSources/bySubject/Overview.html
WebCrawler Searching	http://webcrawler.com/
Welcome to Pathfinder	http://pathfinder.com/@@Y4YH6aFmIgMAQPBa/pathfinder/welcome.html
WhoWhere? PeopleSearch	http://www.whowhere.com/
WWW File Library	http://www.axes.co.jp/~hisashin/lib/index.html
WWW Virtual Library: Geography	http://hpb1.hwc.ca:10002/WWW_VL_Geography.html

Source	Web address
Yahoo	http://www.yahoo.com/
Zip2 Main Search	http://www.zip2.com/

Societies

Source	Web address
AAG	http://www.aag.org/index.html
AAG GIS Specialty Group	http://www.gis.sc.edu/gis/aaggis.html
American Congress on Surveying and Mapping	http://www.landsurveyor.com/acsm/
ASPRS	http://www.us.net/asprs/
Canadian Map Libraries & Archives	http://www.sscl.uwo.ca/assoc/acml/acmla.html
Cartographica	http://utl1.library.utoronto.ca/www/utpress/journal/jour5/car_lev5.htm
Cartography Specialty Group	http://www.csun.edu/~hfgeg003/csg/
Friends of the Earth	http://www.foe.co.uk/index.html
ICA: U.S. National Committee	http://www.gis.psu.edu/ica/ICAusnc.html
Map Societies Around the World	http://www.csuohio.edu/CUT/MapSoc/Index.htm
MapHist Discussion Group	http://kartoserver.frw.ruu.nl/HTML/STAFF/krogt/maphist.htm
NACIS Sixteenth Annual Meeting	http://maps.unomaha.edu/NACIS/Conference.html
NCGE Home Page	http://multimedia2.freac.fsu.edu/ncge/
Society of Cartographers	http://www.shef.ac.uk/uni/projects/sc/
Transactions of the Institute of British Geographers	http://ppt.geog.qmw.ac.uk/
URISA	http://www.urisa.org/

Weather

Source	Web address
Blue Skies Java	http://cirrus.sprl.umich.edu/javaweather/
Current Weather Map	http://www.mit.edu:8001/usa.html
Current Weather Maps/Movies	http://wxweb.msu.edu/weather/
EarthWatch Weather on Demand	http://www.earthwatch.com/
Intellicast	http://www.intellicast.com/
Interactive Weather Browser	http://wxweb.msu.edu/weather/interactive.html

Source	Web address
Interactive Weather Information Network	http://iwin.nws.noaa.gov/iwin/graphicsversion/main.html
Interactive weather maps	http://wxp.atms.purdue.edu/interact.html
Live Access to Climate Data	http://ferret.wrc.noaa.gov/fbin/climate_server
Real-Time Weather Data: Surface Page	http://http.rap.ucar.edu/weather/surface.html
Surface Data Details	http://wxp.atms.purdue.edu/surface_det.html#depict
Texas A&M Meteorology Weather Center	http://www.met.tamu.edu/weather.shtml
The Weather Visualizer	http://covis.atmos.uiuc.edu/covis/visualizer/
Tornadoes	http://cc.usu.edu/~kforsyth/Tornado.html
TV Weather Dot Com- The Weather Supersite	http://www.tvweather.com/
Weather Channel	http://www.weather.com/
Weather Channel - Teachers' Resources	http://www.weather.com/weather_whys/teachers_resources/
WeatherNet	http://cirrus.sprl.umich.edu/wxnet/
World Meteorological Organization	http://www.wmo.ch/

APPENDIX C

MapInfo Product Line Overview

MapInfo Corporation and its worldwide partners provide products and services to meet diverse geographic information needs. Below is a summary list of MapInfo's product line.

- Software products
- Data products
- Data enabling products (self-contained, easily integrated component software)
- Mapping solutions (applications from MapInfo worldwide partners)
- Telecommunication solutions
- Consulting services

- Developer services (tools, technical assistance, and training for developers)
- Training

The following table summarizes major MapInfo products, tools, and solutions, as well as the environments for which they are intended. Each product is described in greater detail in the next section. Contact MapInfo Corporation at the address and phone number found in Appendix D, "Contact Information."

	Desktop	Client/ server	Internet/ intranet	Database tool	Developer tool	Address matching
DataBlade Module						X
MapInfo Desktop	X					X
MapInfo Professional	X	X				X
MapInfo for Power Mac	X					X
MapBasic					X	
MapXtreme		X	X		X	
MapX					X	
MapXsite			X		X	
SpatialWare		X		X		
MapMarker	X	X				X
Microsoft Map	X					

Product Information

Desktop

Products: MapInfo Desktop, MapInfo Professional, MapInfo for Power Mac, MapMarker, Microsoft Map

These mapping software products developed for the desktop are ready to use out of the box. MapInfo Professional and MapInfo for PowerMac are more robust, allowing you to edit tabular and map data, create new map regions, and connect to server based relational databases. They can also be customized. Microsoft Map, an embedded feature in Excel for Windows 95 and Windows 97, allows you visualize data contained in an Excel spreadsheet.

MapInfo Desktop

Released in February 1996, MapInfo Desktop was designed to address an emerging market segment—business users accustomed to office productivity tools (e.g., spreadsheets and databases)—to enable them to take advantage of geographic data analysis in a standalone setting. MapInfo Desktop allows users to display data on a map for presentations or printing. Once data are in MapInfo Desktop and geocoded, users can write simple queries for basic analysis, create thematic maps, change the project, label and annotate, drag and drop a map into an OLE container application, and more. The software can accommodate files created with dBASE, FoxBase, delimited ASCII, Lotus 1-2-3, Microsoft Excel, and other formats.

MapInfo Professional

More powerful than MapInfo Desktop, MapInfo Professional represents a robust, intuitive mapping solution for corporate problem solving. MapInfo Professional offers the highest level of functionality of the MapInfo products designed for business mapping. Appearing below is a partial list of functionality found in MapInfo Professional above and beyond MapInfo Desktop capabilities.

- Run custom MapBasic applications
- Create maps (digitize) and edit map objects extensively
- Read/write remote data directly from Oracle, SyBase, or other relational databases
- Redistrict data
- Perform and save complex queries (such as those requiring table joins)
- Import or export CAD (.dxf), DWG, native ARC/INFO, DGN files
- Connect with GPS software
- Report wizard

MapInfo for Power Mac

MapInfo for Power Mac is the version of MapInfo Professional developed for Power Mac. It offers geocoding, thematic mapping, database queries, and table manipulations for the Mac platform.

MapMarker

MapMarker is a geocoding solution that assigns map coordinates to data records based on address information. MapMarker can be implemented on the client or server side, as a standalone solution, or as part of another application, and it can geocode entire databases in batch mode or interactively as records are entered.

Microsoft Map

Microsoft Map is the result of the partnership between Microsoft and MapInfo. Embedded in Excel for Windows 95 and 97, Microsoft Map allows the user to create a map by simply selecting a range of cells within an Excel worksheet that contain a geographic reference. The user can then modify the map displayed to highlight trends, patterns, and relationships among the data. Because Microsoft Map is an OLE component, it can also be used in a host of other products, such as Microsoft Word, PowerPoint, and even e-mail.

Client/Server

Products: MapInfo Professional, MapXtreme, SpatialWare Workgroup, MapMarker

MapInfo Professional and the SpatialWare/Professional tandem solution are Windows based client mapping solutions with server-side management of map data. MapXtreme offers a client-side browser interface with server-side management of applications and data. SpatialWare provides an intuitive client-side visualization tool for the data warehouse.

MapInfo Professional

See preceding description.

MapXtreme

MapInfo MapXtreme is a mapping application server for company intranets or extranets that developers can quickly and inexpensively implement. MapXtreme offers powerful functionality to support applications such as marketing and sales, inventory management, and product transportation—all cost effectively deployed over a company intranet. Applications running on managed server networks offer lower hardware and administrative costs and dramatically improve application performance, reliability, and security. Because MapXtreme leverages the ease of use and efficiency of Internet technology, companies who once found mapping cost prohibitive can now offer it throughout their organizations at a lower cost per user than in the past.

SpatialWare Workgroup

SpatialWare is a suite of server software technologies for complete spatial information management in a relational database management system setting. At present, SpatialWare supports Oracle and Informix databases. SpatialWare combines business and spatial data in the same system, which maintains central control, data currency, and still permits broad access.

SpatialWare includes an easy to implement spatial addition that leverages Oracle and Informix capabilities; true spatial operation in SQL-based standards; tools to build custom Windows or UNIX clients that provide end-user access to SpatialWare server capabilities; open system, standards based implementation that lowers the cost of integrating with existing information systems; and very large database support.

MapMarker
See preceding description.

Internet/Intranet
Products: MapXtreme, MapXsite

MapXtreme offers a complete Internet/intranet mapping solution; MapXsite is a powerful and flexible "find nearest" application.

MapXtreme
See preceding description.

MapXsite
MapXsite allows Internet developers to integrate "find nearest" mapping functionality into Web sites. Product inventory data can also be accessed, with additional enhancement. MapXsite is fully scalable, the development environment can customize applications on any Windows 95 or Windows NT system, and the product is compatible with popular Web server software.

Database Tools

Product: SpatialWare, SpatialWare DataBlade Module, SpatialWare Workgroup

SpatialWare

See preceding "SpatialWare Workgroup" description.

SpatialWare DataBlade Module

MapInfo's SpatialWare is the world's first and most comprehensive object relational implementation of a geospatial database management system. It is powerful server technology that stores and manages complex spatial data in corporate relational database servers for spatial SQL based querying. SpatialWare is the most advanced server technology for extending the power of databases through spatial analysis. It implements the emerging SQL III Spatial command set for spatial data access, analysis, and modeling in an open, standards based environment.

SpatialWare Workgroup

See preceding description.

Developer Tools

Products: MapBasic, MapXtreme, MapX, MapXsite

MapBasic is a proprietary programming language similar to Basic for creating mapping applications. MapX is an OCX component requiring standard programming languages. MapXsite, described previously, is a prepackaged dealer locator solution for web developers to add mapping functionality to a Web site.

MapBasic

MapBasic enables users to control and extend the functionality of MapInfo Professional through custom programming. The development environment contains a text editor, compiler, linker, and online help. The MapBasic language resembles most popular, structured programming languages and retains the English-like flavor and ease of use of Basic. Users familiar with Visual Basic, Qbasic, or other Basic derived languages will find programming with MapBasic similar. Three categories of applications exist in MapBasic: turn-key applications (complete solutions for a specific need), utilities (e.g., automating tedious tasks), and extensions (e.g., adding new operations to MapInfo Professional).

MapXtreme

See preceding description.

MapX

For application developers, MapX represents the most exciting advance in MapInfo's product line in several years. An OCX component that allows developers to embed a MapInfo map in the application they are writing, MapX enables developers to add maps in applications based in Visual Basic, PowerBuilder, Visual C++, Delphi, and so forth. With MapX, developers work in familiar environments, and mapping becomes part of a larger business application rather than the focus of the application itself.

Address Matching

Products: MapInfo Geocoding DataBlade Module, MapInfo Desktop, MapInfo Professional, MapInfo for Power Mac, MapMarker

MapInfo Geocoding DataBlade Module

The MapInfo Geocoding DataBlade module performs all the necessary calculations to geocode data records stored in an INFORMIX-Universal Server. This DataBlade module encapsulates all functions necessary for creating a geocoding application, maintaining geocoded locations as they are changed or entered in a database, and making data available for visualization and analysis in a client application such as MapInfo Professional. It also handles misspellings, inaccuracies, and other errors in addresses.

MapInfo Desktop

See preceding description.

MapInfo Professional

See preceding description.

MapInfo for Power Mac

See preceding description.

MapMarker

See preceding description.

Appendix D

Contact Information

Blue Marble Geographics
261 Water Street
Gardiner, ME 04345 USA
(800) 616-2725 US and Canada
(207) 582-6747 International
(207) 582-7001 fax
Web site: *www.bluemarblegeo.com*

Bureau of Transportation Statistics
Attn: Customer Service Program Manager
400 7th Street, SW, Room 3430
Washington, DC 20590
(202) 366-3282
Web site: *www.bts.gov*
E-mail: *comments@bts.gov; assistance@bts.gov*

Canada Post Corporation
Retail Analysis Department
2701 Riverside Drive, Suite N0511
Ottawa, ON K1A 0B1
Canada

Claritas Inc.
1525 Wilson Boulevard, Suite 1000
Arlington, VA 22209-2411
(703) 812-2700
(703) 812-2701 fax
Web site: *www.claritas.com*
E-mail: *info@claritas.com*

DeskMap Systems
P.O. Box 40085
Georgetown, TX 78628
(512) 863-6885

Emergency Medical Service Institute
Attn: Christopher Price
4240 Greensburg Pike
Pittsburgh, PA 15221
(412) 351-6604
(412) 351-6616 fax

Dr. Dennis Fitzsimons
Department of Geography & Planning
Southwest Texas State University
San Marcos, TX 78666-4616
E-mail: *df02@swt.edu*

Contact Information

IntelleVue
Attn: Jeffrey Davis or Angela Whitener
P.O. Box 470142
Tulsa, OK 74147-0142
(918) 250-5561
(918) 254-7245 fax
Web site: *www.intellevue.com*
E-mail: *Jdavis@intellevue.com*,
　　　　Awhitener@intellevue.com

Louis Dreyfus Natural Gas
Attn: Stan Nickel, Systems Engineer
14000 Quail Springs Parkway
Oklahoma City, OK 73134-2600
(405) 740-5274
(405) 751-5129 fax
Web site: *www.ldng.com*

MapInfo Corporation
One Global View
Troy, NY 12180-8399 USA
(800) 488-3552 USA
(518) 285-7110 International
(518) 285-6070 fax
Web site: *www.mapinfo.com*
E-mail: *sales@mapinfo.com*

MISA - Market Information Services of America, Inc.
121 South Wilke Road, Suite 300
Arlington Heights, IL 60005
(800) 685-MISA
(800) 286-7995 fax
Web site: *www.misa.com*
E-mail: *webmastr@misa.com*

Northwood Geoscience Ltd.
89 Auriga Drive
Nepean, ON K2E 7Z2
Canada
(613) 224-2020
(613) 224-2785 fax
Web site: *northwoodgeo.com*

On Target Mapping
1051 Brinton Road
Pittsburgh, PA 15221
(800) 700-6277 (MAPS)
(412) 241-7622
(412) 241-7709 fax
Web site: *www.otmapping.com*
E-mail: *Sales@otmapping.com*

Public Safety Associates, Inc.
Attn: Rick Peters
2639 Walnut Lane, Suite 113
Dallas, TX 75229
(214) 956-0911
(214) 956-8816 fax
Web site: *www.psadallas.com*
E-mail: *sales@psadallas.com*

Contact Information

Rural Press Ltd.
Raleigh Park
12 Todman Avenue
Kensington, NSW 2033
Australia
61 (2) 9313 8444
61 (2) 9663 2322 fax
Web site: *www.rpl.com.au/index.html*

Thrifty Rent-A-Car System, Inc.
Attn: Sandy Carter, Manager
5330 East 31st Street, CIMS 1092
Tulsa, OK 74153-0250
(918) 665-3930
(918) 669-2213 fax
E-mail: *sandy.carter@thrifty.com*

Visimark, LLC
2100 Clearwater
Oak Brook, IL 60521
Web site: *www.Visimark.com*

Index

Numerics

9-section plat map
   described 216
   oil and gas well data 219–220
   required data 218–220
   section/township/range data 218–219

A

ABS
   CDATA MapInfo add-on 173
   CDATA96, innovations described 177–179
accounting information systems applications 109
advertising
   banking industry, direct mail campaigns with MapInfo 118–119
   MapInfo applications 23–31
   Rural Press/CDATA case study 171–179
area map, defined, petroleum price change study 155
attractiveness grid, Canadian Business Information (CBI) data 185
attribute data, defined 13
Australian Bureau of Statistics. *See* ABS

B

banking
   advertising, direct mail campaigns with MapInfo 118–119
   business conditions today 113
   changing customer base, effects of 114
   CRA compliance
      desktop mapping solutions 117–118
      history of 116–117
   GIS and desktop mapping, effects of 114
   marketing, MapInfo applications 118–119
   network optimization
      benefits of site selection and analysis 115–116
      importance of customer service 115
      key market elements listed 115
Blue Marble Geographics, Inc., Geographic
   Tracker software 85–87
business data
   geographic component in 9–11
   mapping, benefits of 2–8

C

Canada Post case study
   defining market area 184
      Forward Sortation Areas 184
   determining customer base 183
   determining potential store revenue 192
   determining relative store strength 191, 192
   determining total revenue potential 184
   establishing optimal store locations 189
   establishing store attractiveness 183–184
   Huff model 192
   introduced 181
   locating existing stores 184
   methodology summarized 182
   patronage probability grid 189–191
   performing gap analysis 189
   store attractiveness grid 185
      standard industry code (SIC) data 185
      Vertical Mapper use 187–188
      weighting retail areas 186–187
Canadian Business Information. *See* CBI data
catastrophe planning and response
   introduced 127
   Mississippi Flood 127
CBI data
   Canada Post case study, analyzing store attractiveness 185
CDATA
   case study, statistical data collection applications 174–177
   CDATA96 application, innovations described 177–179
   described 173
Census data
   mapping 2–8

PRIZM lifestyle segmentation data 24–27
TIGER/Line files 33
Charting Pro 31
ChartLink 31
Community Reinvestment Act. *See* CRA
competition
    evaluating, site selection and analysis 46–47
Consumer Expenditure Survey 35–36
consumer spending data 35–36
    market segmentation and analysis 34–36
CRA
    banking industry compliance
        desktop mapping solutions 117–118
        history of 116
    custom applications 118
    MapInfo CRA function 118
customer analysis
    banking industry, direct mail campaigns with MapInfo 118–119
customer migration
    analyzing, petroleum price change study 157–158
customer service
    importance in banking industry 115–116
    MapInfo Professional applications 36–40
        "find nearest" functionality 38–40
        locational functionality 38–40
    telecommunications, GIS applications in 99–101

D

data
    analysis, level of geographic detail 11
    attribute, defined 13
    business
        benefits of mapping 1–8
        geographic component in 9–11
    consumer spending
        Consumer Expenditure Survey 35–36
        market segmentation and analysis 34–36
        site selection and analysis 41–46
    feature, defined 13
    geographic
        address geocoding 32–33
        and business data 9–11
        geographic reference, defined 9–10
    lifestyle, market segmentation applications 24–28
    MapInfo data sources
        client data 132
        data partner products 133
        third party data 133
    telecommunications 105
    use in insurance underwriting 121
decision sciences
    GPS technology
        Geographic Tracker software 85–87
        introduced 78–82
        Mobiletrack Pty Limited 82
        newspaper circulation management 83
        U.S. Postal Service delivery management 84
        uses 87
    introduced 57
    inventory management
        beverage company deliveries 58–60
        discussed 57–66
        parking lot inventory management 63
        supermarket inventory management 61–62
        supermarket product placement 61–62
        warehouse inventory management 63–66
    routing logistics
        package pickup and delivery 89–90
        tracking utility repairs 88
    transportation industry
        mass transit 75–76
        railroad management 66–73
        roadway management 73–75
demographic data
    data variables described 45
    site selection and analysis 41–46
        projecting change with thematic maps 44–
        projecting demographic data 43–45
DemoTrack 31
desktop mapping. *See* mapping
DRIVE and DRIVE PLUS 195–198

E

Emergency Medical Service Institute. *See* EMSI
emergency response
    industry use of railroad data 71
EMSI case study
    business mission 194
    calculating response time 195–197
    introduced 193–194
    license inspections 198
environmental risk assessment
    insurance industry 121
        available historical environmental data 121

F

feature data, defined 13
fundamental demographic data, described 45

Index

G

gap analysis
    Canada Post case study 189
    real estate management 135
geocoding
    address, explained 32
        improving success rate 32–33
    competitor locations, site selection and analysis 46–47
    TIGER/Line files 33
geographic data
    address geocoding, explained 32
        improving success rate 32–33
    and business data 9–11
    geographic reference, defined 9–10
    level of geographic detail 11
    TIGER/Line files 33
geographic information systems. *See* GIS
geographic reference, defined 9–11
Geographic Tracker, described 85, 86
GIS
    and desktop mapping, introduced 12–14
    effects on banking industry 114
    feature data, defined 13
    history of 14
    implementing
        choosing a consultant 122–123
        choosing a software package 122–123
    importance of data 132–133
    and railroad management 68
    real estate, GIS home search applications 137
    role in telecommunications 96–101
Global Positioning Systems. *See* GPS technology
GPS technology
    causes of signal degradation 78
    common uses 78
    Geographic Tracker software 85–87
    GPS solutions, power company example 79
        Westinghouse Hanford Company example 80
    introduced 78–82
    Mobiletrack Pty Limited 82
    newspaper circulation management 83
    receivers
        mapping grade 81
        navigation grade 81
        survey grade 81
    U.S. Postal Service delivery management 84

H

Huff model
    Vertical Mapper, Canada Post case study 189–191, 192
human resources management
    MapInfo information systems applications 110

I

information systems
    EMSI case study 193–199
    Louis Dreyfus Natural Gas case study 213–227
    MapInfo applications 105
        accounting, management and reporting 109
        human resources management 110
        Internet applications 106
        oil and gas well mapping 108
        political data management 110
insurance
    business mapping applications 120
    catastrophe planning and response 127–128
        Mississippi Flood 127
    recent history 120
    risk concentration analysis 124–126
        On Target Mapping databases 125
        probable maximum loss analysis 124
    sales and planning
        desktop mapping applications, health care 128
        target marketing 130
    underwriting
        calculating auto drive distances 121
        environmental risk assessment 121
            available historical environmental data 121
        introduced 120
IntelleVue 144
    MapInfo application, Louis Dreyfus Natural Gas case study 221–227
    oil and gas well mapping with MapInfo 108
intermodal transportation 72
Internet
    information systems, MapInfo applications 106
    PageNet case study 201–211
inventory management
    beverage company deliveries 58–60
        product placement 58–60
    introduced 57
    parking lot 63
    supermarket 61–62
        product placement 61–62
    warehouse 63–66
inverse distance weighting (IDW)
    Canada Post case study 187–188

L

lifestyle segmentation
    application to site selection 29
    banking applications 28
    product target markteting 30
Louis Dreyfus Natural Gas 221–227
    IntelleVue's MapInfo application 221–227
    introduced 213–215
    MapInfo application goals, defined 215–216
    9-section plat map
        described 216
        oil and gas well data 219–220
        required data 218–220
        section/township/range data 218–219
    other MapInfo applications 227
    Petroleum Information (PI)/Dwights, described 220

M

MapChart 31
MapInfo
    add-on software 31
    commands 18
    U.S. Census data, mapping 5–8
MapInfo Corporation
    description 14–15
    and history of GIS 14
    MapMarker 33
    partnership with Microsoft 15
    Web address 118
MapInfo Professional applications
    address geocoding 32–33
    advertising 23–31
    banking industry
        CRA function 118
        direct mail campaigns with 118–119
    CDATA application described, Rural Press/CDATA case study 173–177
    CRA function 118
    customer service 36–40
    data sources
        client data 132
        MapInfo data partner products 133
        third party data 133
    "find nearest" functionality 38–40
    Geographic Tracker software 85–87
    GPS technology and Mobiletrack Pty Limited 82
    information systems 105–111
        accounting management and reporting 109
        Internet applications 106
        oil and gas well mapping 108
        political data management 110
    IntelleVue's Louis Dreyfus Natural Gas application 221–227
    inventory management 57–66
        beverage company delivery 58–60
        managing product data 60
        parking lot 63
        supermarket 61–62
        supermarket product placement 61–62
        warehouse 63–66
locational functionality 36–40
Louis Dreyfus Natural Gas case study 227
MapLink software, Visimark case study 234–237
MapMarker, real estate applications 134
marketing 23–31
markets, described 16
mass transit management 75–76
petroleum price change study
    basic user interface characteristics 146–147
    credit card customer area map 154–155
    credit card customer point map 151–153
    customer map parameters, described 155–156
    described 144
    map customer migration, described 156–157
    payment method comparison 160
    premium payment method comparison 159
    Pricing Analysis menu 149–151
    Site Map menu 148
real estate industry 134–137
redistricting 52–55
    advertising regions 54
    police beats 54
routing logistics
    package pickup and delivery 89–90
    school bus routing 92–94
sales 23–31
site analysis modeling 49–51
site selection and analysis 41–51
    "backing in" technique 45
telecommunications
    competitive analysis 104
    customer service 100
    developing rural phone networks 102
    tracking buried cables 103
Thrifty Rent-A-Car case study
    customer origination map analysis 167–168
    data collection techniques 163–165
    effects of MapInfo use 168
    generating customer origination maps 165–

Index

167
   MapInfo use 163
tracking utility repairs 88
trade area analysis 48–49
transportation industry
   intermodal transportation 72–73
   railroad management 71
      and telecommunications industry 71
      emergency response organizations 71
      GIS solution, described 68
      profitability analysis 69–71
      U.S. Railroad Database, described 66–67
   roadway construction, management, and operation 74
      planning 73
      traffic counting 73
MapLink software
   development of, history 234–235
   geocoding function 235
   mapping selected property set 236–237
   searching by map location 236
MapMarker
   address geocoding 33
   real estate applications 134
MapMarket 110
mapping
   business data, geographic component in 9–11
   and GIS technology, history 14
   and GIS technology, introduced 12–14
   market segmentation 24
   as persuasive tool 1–8
   U.S. Census data 2–8
market segmentation
   consumer spending data 34–36
      Consumer Expenditure Survey 34–36
   defined 24
   lifestyle data 24–28
      PRIZM system 24–27
marketing
   banking industry, direct mail campaigns with MapInfo 118–119
   maintaining current customers with GIS 98
   MapInfo Professional applications 23–31
   market segmentation. *See* market segmentation
   PageNet case study 201–211
   petroleum price change study 143–161
   telecommunications, attracting new customers with GIS 97–98
   Thrifty Rent-A-Car case study 163–169
mass transit

management 75–76
   MapInfo Professional applications 75–76
MISA 31
Mobiletrack Pty Limited, use of GPS technology 82
modeling
   Canada Post case study 181–192
   site selection and analysis 49–51
      choosing modeling variables 50–51
      predictive model 50
      role of demographic data 50–51
      screening model 50

O

oil and gas wells
   mapping with MapInfo 108
On Target Mapping
   DRIVE and DRIVE PLUS software 195–198
   environmental and risk analysis databases 125–126

P

PageNet case study
   introduced 201–202
   PCS Site Manager
      application goals, defined 202
      described 202–204
      future enhancements 210–211
      marketing/executive level overview 205
      Real Time Site Analysis Probe 207–210
      region analysis options 204
   Public Safety Associates, Inc. 201–202
Panel View, add-on software 31
PCS Site Manager. *See* PageNet case study
Petroleum Information (PI)/Dwights, described 220
petroleum price change study. *See* MapInfo applications
point map, petroleum price change study 152
politics, MapInfo applications 110
PRIZM, lifestyle segmentation data 24–27
probable maximum loss analysis, insurance industry 124
product placement, inventory management
   beverage company 60
   supermarket 61–62
product target marketing 30
profitability analysis, in railroad management 69–71

R

railroad management
  intermodal transportation 72–73
  profitability analysis 69–71
  and telecommunications industry
    emergency response 71
  U.S. Railroad Database, described 66–67
real estate
  common desktop applications 131
  GIS home search applications 137
  management 135–136
    gap analysis 135
  site selection 134–135
  Visimark case study 229–237
redistricting
  advertising regions 54
  introduced 52
  police beats 54
  uses 52–55
retail
  petroleum price change study 143–161
  site analysis, Canada Post case study 181–192
risk concentration analysis 124–126
  insurance industry
    On Target Mapping databases 125
    probable maximum loss analysis 124
roadway management
  construction and operation 74
  traffic counting 73
  transportation planning 73
routing
  EMSI case study 193–199
routing logistics
  package pickup and delivery
    best route calculation 89
    routing software, described 89–91
    shortest path calculation 89
  school bus routing 92–94
Rural Press/CDATA case study
  CDATA applications, statistical data collection 174–177
  CDATA96 innovations 177–179
  Rural Press Limited 171–172

S

sales
  insurance, desktop mapping applications 128–131
  MapInfo Professional applications 23–31
  Rural Press/CDATA case study 171–179

SIC data
  Canada Post case study 185
site selection and analysis
  "backing in" 45
  banking industry network optimization 115–116
  demographic variables 45
  evaluating demographic data 41–46
    "snake" graph 42
  identifying competition 46–47
  introduced 41
  modeling. *See* modeling
  projecting demographic data 44–45
  real estate 134–135
  trade area analysis 48–49
    drive distance 48–49
    drive time 48–49
    ring 48–49
  uses 41
  Visimark case study 229–237
standard industry code data. *See* SIC data
Surround View, add-on software 31

T

target marketing
  billboard advertising 31
  market segmentation 24–28
telecommunications
  and railroad industry 71
  competitive analysis 104
  customer service
    GIS applications 99–101
  data 105
  effects of deregulation 96, 102, 104
  introduced 95–97
  marketing and sales
    attracting new customers with GIS 97–98
    maintaining current customers with GIS 98
  network management and planning 101–102
    rural access 102
  PageNet case study 201–211
  role of GIS 96–97
  tracking buried cables 103
thematic map, projecting demographic change with 44
Thrifty Rent-A-Car case study. *See* MapInfo applications
TIGER/Line files
  address geocoding 33
  history 14

Index

trade area, defined 48
trade area analysis
   drive distance 48–49
   drive time 48–49
   ring 48–49

U

U.S. Postal Service
   delivery management with GPS technology 84
U.S. Railroad Database, described 66–67
underwriting. *See* insurance industry

V

Vertical Mapper, Canada Post case study 181–192
   determining relative store strength 191
   Huff model function 189–191
   patronage probability grid 189–191
   retail density grid 187–188
      inverse distance weighting (IDW) 187–188
Visimark case study
   MapLink software
      geocoding function 235
      mapping selected property set 236–237
      searching by map location 236
   Visimark LLC software described 229–231
      integrating with GIS functionality 233–235
      resource savings 233
      searching and database capabilities 231–232

W

Westinghouse Hanford Company
   use of GPS technology 80

More OnWord Press Titles

NOTE: All prices are subject to change.

Computing/Business

Lotus Notes for Web Workgroups
$34.95

Mapping with Microsoft Office
$29.95 Includes Disk

The Tightwad's Guide to Free Email and Other Cool Internet Stuff
$19.95

Geographic Information Systems (GIS)

GIS: A Visual Approach
$39.95

The GIS Book, 4E
$39.95

GIS Online: Information Retrieval, Mapping, and the Internet
$49.95

INSIDE MapInfo Professional
$49.95 Includes CD-ROM

Minding Your Business with MapInfo
$49.95

MapBasic Developer's Guide
$49.95 Includes Disk

Raster Imagery in Geographic Information Systems
$59.95 Includes color inserts

INSIDE ArcView GIS, 2E
$44.95 Includes CD-ROM

ArcView GIS Exercise Book, 2E
$49.95 Includes CD-ROM

ArcView GIS/Avenue Developer's Guide, 2E
$49.95 Includes Disk

ArcView GIS/Avenue Programmer's Reference, 2E
$49.95

ArcView GIS /Avenue Scripts: The Disk, 2E
Disk $99.00

ARC/INFO Quick Reference
$24.95

INSIDE ARC/INFO, Revised Edition
$59.95 Includes CD-ROM

Exploring Spatial Analysis in Geographic Information Systems
$49.95

Processing Digital Images in GIS: A Tutorial for ArcView and ARC/INFO
$49.95

Softdesk

INSIDE Softdesk Architectural
$49.95 Includes Disk

Softdesk Architecture 1 Certified Courseware
$34.95 Includes CD-ROM

Softdesk Architecture 2 Certified Courseware
$34.95 Includes CD-ROM

INSIDE Softdesk Civil
$49.95 Includes Disk

Softdesk Civil 1 Certified Courseware
$34.95 Includes CD-ROM

Softdesk Civil 2 Certified Courseware
$34.95 Includes CD-ROM

MicroStation

INSIDE MicroStation 95, 4E
$39.95 Includes Disk

MicroStation 95 Exercise Book
$39.95 Includes Disk
Optional Instructor's Guide $14.95

MicroStation 95 Quick Reference
$24.95

MicroStation 95 Productivity Book
$49.95

Adventures in MicroStation 3D
$49.95 Includes CD-ROM

MicroStation for AutoCAD Users, 2E
$34.95

MicroStation Exercise Book 5.X
$34.95 Includes Disk
Optional Instructor's Guide $14.95

MicroStation Reference Guide 5.X
$18.95

101 MDL Commands (5.X and 95)
Executable Disk $101.00
Source Disks (6) $259.95

CATIA

INSIDE CATIA
$80.00 Includes CD-ROM

CATIA Reference Guide
$49.95

Other CAD

Fallingwater in 3D Studio
$39.95 Includes Disk

SunSoft Solaris

Solaris 2.X for Managers and Administrators Guide, 2E
$34.95

SunSoft Solaris 2.* User's Guide
$29.95 Includes Disk

SunSoft Solaris 2.* Quick Reference
$18.95

Five Steps to SunSoft Solaris 2.*
$24.95 Includes Disk

SunSoft Solaris 2.* for Windows Users
$24.95

Windows NT

Windows NT for the Technical Professional
$39.95

HP-UX

HP-UX User's Guide
$29.95

Five Steps to HP-UX
$24.95 Includes Disk

OnWord Press Distribution

OnWord Press books are available worldwide from OnWord Press and your local bookseller. For order information, terms, or listings of local booksellers carrying OnWord Press books, call toll-free 1-800-4-ONWORD (1-800-466-9673) or 505-474-5130; fax 505-474-5030; write to OnWord Press, 2530 Camino Entrada, Santa Fe, New Mexico 87505-4835, USA, or e-mail orders@hmp.com. OnWord Press is a division of High Mountain Press.

Comments and Corrections

Your comments can help us make better products. If you find an error, or have a comment or a query for the authors, please contact us at the address below, send e-mail to cleyba@hmp.com, or call us at 1-800-466-9673.

OnWord Press, 2530 Camino Entrada, Santa Fe, NM 87505-4835 USA

On the Internet: http://www.hmp.com